FLOWERS
OF THE
SOUTHWEST DESERTS

By Natt N. Dodge

Illustrated by Jeanne R. Janish

Editor: Earl Jackson

**SOUTHWEST PARKS AND
MONUMENTS ASSOCIATION**
339 South Broad Street
Box 1562, Globe, Arizona 85501

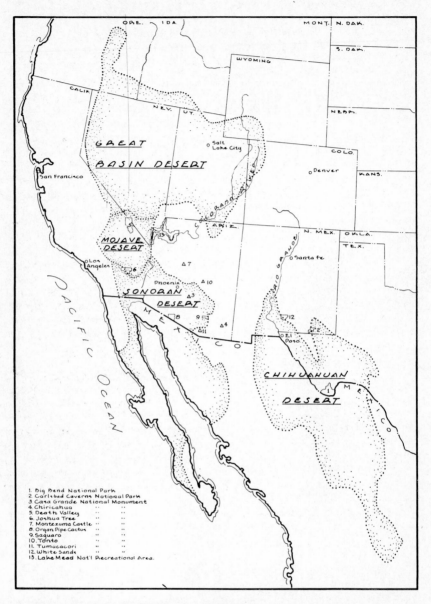

1. Big Bend National Park
2. Carlsbad Caverns National Park
3. Casa Grande National Monument
4. Chiricahua   "   "
5. Death Valley   "   "
6. Joshua Tree   "   "
7. Montezuma Castle   "   "
8. Organ Pipe Cactus "   "
9. Saguaro   "   "
10. Tonto   "   "
11. Tumacacori   "   "
12. White Sands   "   "
13. Lake Mead Nat'l Recreational Area.

    Desert Areas of the West—this booklet deals with the common plants of three of them: (1) the Chihuahuan; (2) the Sonoran; and (3) the Mojave.

    Plants of the higher plateau country of from 4,500 to 7,000-feet elevation are shown and described in "Flowers of the Southwest Mesas," companion volume to this one, by Pauline M. Patraw and Jeanne R. Janish.

    Mountain zone vegetation (from the Ponderosa Pine belt, or about 7,000 feet, on up) is the subject of "Flowers of the Southwest Mountains," the third of the triad, by Leslie P. Arnberger and Jeanne R. Janish.

FLOWERS of the SOUTHWEST DESERTS

HOW TO USE THIS BOOKLET

To get full value from this booklet, it is important that you know how to make the greatest use of it. The purposes of the booklet are: (1) to introduce the common desert flowers to newcomers to the Southwest; and (2) to give some background information about the plants' interesting habits and how they have been and are used by animals, by the native peoples, and by the settlers. Every effort has been made to present accurate, if not always complete, information.

Since there are more than 3,200 plants recorded from Arizona alone, and this booklet attempts to introduce you to the common plants of desert areas in Texas, New Mexico, and California in addition to Arizona, it is apparent that you will find an enormous number of flowers which are not included. Therefore, a painstaking effort has been made to select the commonest or most spectacular; that is, those which you will naturally stop to look at and say, "Who are you?"

For ease in identification, we have arranged the flowers according to color of their petals. When you meet a flower to whom you would like an introduction, first note the color of its petals. Don't jump too quickly to a conclusion, for what at first glance may seem to be pink, careful examination may prove to be lavender, violet, or purple. Once you feel reasonably sure of the color, turn to the section in which flowers of that color are listed and examine the sketches. Find something that looks similar?

Now check the size of the plant as indicated in the sketch and text. Does the text list the flower as occurring in the particular desert area (see map on opposite page) where you are? Is the blossoming season correct? Do other details check? If so, the chances are that you have the right flower—or at least a close relative. Close enough, anyway, so that you may be reasonably safe in calling the flower by its common name. Of course if a botanist happens along, he may point out that you have *Penstemon parryi* whereas you thought you had struck up an acquaintance with *Penstemon pseudospectabilis*. However, it's a penstemon, even tho' a sister of the one you thought you were meeting. Perhaps you'll run across a dozen other brothers and sisters before you happen onto the member of the genus common enough to be listed specifically in our Desert Who's Who.

Certain of the desert flowers change color with age. Also, during off seasons, some of the really common flowers don't show up in large numbers while a few of the rarer ones may take their turn at brightening up the desert. Furthermore, in a few cases such as the oleander, the species comes in two colors, red flowers on one plant and white on another. The *poinciana*, or "bird-of-paradise," flower has yellow petals, but the rest of the flower is red, so it's a toss-up which color you might call it. The beavertail pricklypear has magenta flowers while its very close relative, the Engelmann pricklypear, has yellow blossoms, yet in this booklet it has been necessary to put them both on the same page in the "yellow" section.

3

So, this booklet makes no claim to perfection, and these discrepancies add certain hazards to the game. You may strike out several times before getting to first base. As you become accustomed to using it, home runs will come more frequently, and you will soon begin to have a lot of fun. If any particular species especially interests you, once you are certain of its identity you can readily find out more about it by following up in one or more of the publications we have listed under the heading "References."

A few of the common desert flowers have been left out of this booklet—purposely. The reason is that, although they are well represented among desert flowers, they are even more common throughout non-desert parts of the Southwest. You will find them in a companion booklet: Pauline M. Patraw's "Flowers of the Southwest Mesas." They belong principally to the following groups: cottonwood, rabbitbrush, snakeweed, saltbush, apache-plume, clematis, skunkbush sumac, blanketflower, sunflower, groundsel, elder, blazing star and morning-glory.

PLANT NAMES

Be Serious About Plant Names—But Not Too Serious

It has often been said that "a rose by any other name would smell as sweet." Although the statement is literally true, we are often disappointed, perhaps offended, when we find some flower friend of long acquaintance called by another, and, to our minds, inferior name. Also, we dislike the attachment of a name which we have long associated with a certain plant to another, and perhaps less attractive, flower.

Common names are by no means standardized in their usage, and a well known plant in one part of the country may be called by an entirely different name somewhere else. Also, certain names are applied to a number of plants which more or less resemble one another. For instance, the name "greasewood" is applied to almost any plant that has oily or highly inflammable leaves; and with the avid reading by eastern people of Zane Grey's and other "westerns," any shrubby plant with grayish foliage covering large areas of western land immediately becomes "sagebrush." This is particularly irritating to inhabitants of the desert areas treated in this booklet because true sagebrush *(Artemisia tridentata)* rarely grows below elevations of 6,000 feet. The loose application of common names is a confusing annoyance to wildflower enthusiasts.

To avoid this confusion and establish a method of naming that will be uniform throughout the world, botanists have developed a system using descriptive Latin names and grouping plants into genera and families based upon their relationships to one another as determined by their physical structure. Unfortunately for the layman, this system is so technical and the Latin names so unintelligible that he becomes completely bewildered. Furthermore, advanced botanical studies result in continued regroupings and changes in names so that the amateur botanist finds it impossible to keep up. Botanists who specialize in plant nomenclature have a tendency to become so involved with the technicalities of naming that their writings bristle with minute descriptions of anatomical details and the reader searches in vain for such basic information as a simple statement of the color of the flowers.

The majority of common flowers have several to many common names. This is particularly true in the Southwest where some plants have names in English, Spanish, and one or more Indian languages. In addition, of course, each species has its scientific name. An effort has been made in this booklet to give as many of the names applied to each selected flower as are readily available. This not only aids in identification, but adds to its interest. The reader then finds himself in the enviable position of being able to scan the field and choose whichever name appeals to him with the reasonable assurance that he is right—at least in one locality.

Since this booklet was written by a layman for the use and enjoyment of other laymen, it violates a number of botanical, or taxonomic, principles. These violations have been committed with no spirit of disrespect, but in an effort to avoid confusion, conserve space, and keep a complicated and involved subject as simple as possible. The writer believes that the visitor to the desert who has a normal pleasure in nature is interested in the flowers because of their beauty and their relationships with other inhabitants of the desert, including mankind.

THE DESERT — WHAT AND WHERE IS IT?

In this booklet we are dealing with DESERT flowers, so it seems logical to take a moment to check up on the desert itself. What is a desert, and how may we recognize one when we see it?

"A desert," stated the late Dr. Forrest Shreve, "is a region of deficient and uncertain rainfall." Where moisture is deficient and uncertain, only such plants survive as are able to endure long periods of extreme drought. Desert vegetation is, therefore, made up of plants which, through various specialized body structures, can survive conditions of severe drought. In general, the deserts of the world are fairly close to the equator, so they occur in climates that are hot as well as dry. Plants in the deserts of the Southwest must endure long periods of heat as well as drought.

In North America, major desert areas are located in the general vicinity of the international boundary between Mexico and the United States. Due to various differences in elevation, climatic conditions, and other factors, certain portions of this Great American Desert favor the growth of plants of certain types. Based on these general vegetative types, some botanists have catalogued the Great American Desert into four divisions, as follows (see map):

1. Chihuahuan Desert: Western Texas, southern New Mexico, and the Mexican states of Chihuahua and Coahuila.
2. Sonoran Desert (Arizona Desert): Baja (Lower) California, northern Sonora, and southern Arizona.
3. Mojave-Colorado Desert (California Desert): Portions of southern California, southern Nevada, and northwestern Arizona.

4. Great Basin Desert: The Great Basin area of Nevada, Utah, and northeastern Arizona.

It is of especial interest to note that certain plants such as creosotebush *(Larrea tridentata)* seem to thrive in several of these desert areas while others are found in great abundance in only one. Plants that grow in profusion in only one desert are spoken of as "indicators" of that particular desert. Any person interested in desert vegetation soon learns the major indications, not only of the different deserts, but of different sections or elevations in the same desert. Here are some of the better-known indicator plants:

1. Chihuahuan Desert: Lechugilla *(Agave lechuguilla);*
2. Sonoran Desert: Saguaro *(Cereus giganteus);*
3. Mojave-Colorado Desert: Joshua-tree *(Yucca brevifolia);*
4. Great Basin Desert: Big sagebrush *(Artemisia tridentata).*

This publication deals with the common plants and flowers of the Chihuahuan, Sonoran, and Mojave-Colorado Deserts. Since these names are strange to many visitors to the Southwest, the writer has taken the liberty of applying descriptive names as synonyms. In this booklet the Chihuahuan Desert is called the Texas Desert, the Sonoran Desert is referred to as the Arizona Desert, and the Colorado-Mojave Desert is considered as the California Desert.

Whenever possible, the desert in which a particular species of plant is most common is indicated; however, this should not be interpreted too rigidly as most of the plants in this book grow in more than one desert and some grow in all.

Because the Great Basin Desert is a region of higher elevation and is influenced by other factors not common to the three portions of the Great American Desert covered in this booklet, its vegetation is more like that of the plateaulands and foothills of the Southwest. Therefore, its flowers are included in a companion booklet, Polly Patraw's "Flowers of the Southwest Mesas."

NATIONAL PARKS AND MONUMENTS AS WILDFLOWER SANCTUARIES

Someone has called National Parks and Monuments "The Crown Jewels of America." A part of their beauty and irreplaceable value is because the more than 200 units of the National Park System which extends from Florida to Alaska and from Hawaii to Maine, are and have been wildflower sanctuaries. Not only do native plants live under natural conditions, but they are protected from picking, from grazing of domestic livestock, and from the competition of exotic species, and from other activities of mankind that would disrupt their normal habitat or disturb their native way of life.

Uniformed employees of the National Park Service feel complimented whenever visitors show an interest in the natural features of the areas they protect, and are happy to assist them in locating rare species or especially beautiful or spectacular specimens. Range and grazing specialists are more and more using the natural vegetation of National Parks and Monuments as "check plots" to aid them in studying ways and means of preserving the level of grazing value on the open ranges.

Within the desert areas of the Southwest there are a number of National Parks and Monuments. Three Monuments (Joshua Tree in California, Organ Pipe Cactus and Saguaro in Arizona) have been created primarily to save from exploitation and destruction outstnding areas of typical desert vegetation. Although the others have been established to protect and preserve geologic, historic, or archeologic values of national significance, they are all wildflower sanctuaries. In California, Death Valley National Monument is outstanding in its variety of desert flowers. Lake Mead National Recreation Area, of which Hoover Dam is the center, has exceptional displays of various forms of desert plants. A great variety of desert vegetation will be shown and, if desired, explained to the interested visitor, by National Park Service rangers at Chiricahua, Tonto, Montezuma Castle, Casa Grande, and Tumacacori National Monuments in Arizona. Of course the really great displays of desert botany and ecology are featured at Organ Pipe Cactus and Saguaro National Monuments.

In New Mexico, Chihuahuan Desert vegetation is particularly abundant at Carlsbad Caverns National Park. A number of desert forms, especially interesting because of the effect upon them of the ever-moving gypsum dunes, are found at White Sands National Monument, near Alamogordo. Another outstanding Chihuahuan Desert wildflower sanctuary is Big Bend National Park in southwestern Texas.

Photography is encouraged in all of the National Parks and Monuments. By asking a ranger, you will be able to learn where the various flower displays may be found, the best time of day to obtain good results, and other suggestions helpful in obtaining photographs of desert wildflowers at their very best.

Each year several western and southwestern magazines, newspapers, television and radio stations present bulletins and other information about moisture and other pertinent conditions in the desert, spotlighting areas in which outstanding wildflower displays are developing, and in some cases suggesting areas where spectacular displays may be expected if the weather follows conventional behavior patterns.

DESERT PLANTS

Many people think of a desert as an area of shifting sand dunes without vegetation except in areas where springs provide moisture. This is by no means true of our Southwestern deserts which are characterized by a rich and diversified plant cover. However, the majority of true desert plants are equipped by Nature to meet conditions of high temperatures and deficient and uncertain precipitation. The way in which desert plants, closely related to common species found growing under normal temperatures and moisture conditions, have adapted themselves to meet the severe requirements of desert life is truly remarkable and forms an absorbing and fascinating study.

Shreve groups desert plants into three categories based on the manner in which they have contrived to conquer the hazards of desert life.

These are: 1. Drought-escaping plants; 2. Drought-evading plants; 3. Drought-resisting plants.

Drought-escaping plants are the "desert quickies," or ephemerals. Taking advantage of the two seasons of rainfall on the desert (midsummer showers and midwinter soakers) they develop rapidly, blossom, and mature their seeds which lie dormant in the soil during the rest of the year, thus escaping the season of heat and drought. There are two groups of these "quickies," the summer ephemerals and the winter ephemerals. The former are hot-weather plants; the latter are species that thrive during the cool, moist weather of winter and early spring. These "quickies" present their spectacular floral displays only following seasons of above-average precipitation.

Drought-evading plants (in common with the deciduous plants of northern and colder climes which remain dormant while below-freezing temperatures prevail), meet the heat and drought by reducing the bodily processes to maintain life only. They drop their leaves, and remain in a state of dormancy until temperature and moisture conditions, suitable to renewed activity, again prevail.

The *drought-resisting* plants are the bold spirits which take the worst the desert has to offer without flinching, or resorting to evasive tactics. Chief among these are the cactuses which store moisture in their spongy stem or root tissues during periods of rainfall, using it sparingly during drought. To reduce moisture loss to a minimum, they have done away with their leaves, the green skin of their stems taking over the function of foliage. Other plants, such as the mesquite, develop deep or widespread root systems that extract all the moisture from a huge soil area. Most drought-resisters either cut down their leaf surface to an irreducible minimum, or coat the leaves with wax or varnish, thus restricting the loss of moisture.

Methods, techniques, devices, or body modifications which desert plants have developed or evolved to enable them to withstand the rigors of long-continued drought and heat are legion. Many of them are known and understood, but it is probable that there are many others which scientists have not yet discovered.

ACKNOWLEDGEMENTS

For numerous helpful suggestions, lists of common flowers, herbarium and fresh specimens for use in preparing illustrations, and for assistance in many other ways, the author and illustrator proffer sincere thanks to the following: Glen Bean, L. Floyd Keller, Walter B. McDougall, William R. Supernaugh, Dr. Norman C. Cooper, Mrs. Robert Gibbs, Leslie N. Goodding, Edmund C. Jaeger. Thomas H. Kearney, Robert Peebles, (who kindly reviewed the manuscript), Paul Ricker, and Barton H. Warnock. Color picture credits: cover, courtesy Dan Peterson. Desert-marigold, enceliopsis, Joshua-tree in flower, Jeanne R. Janish. All others, Natt N. Dodge.

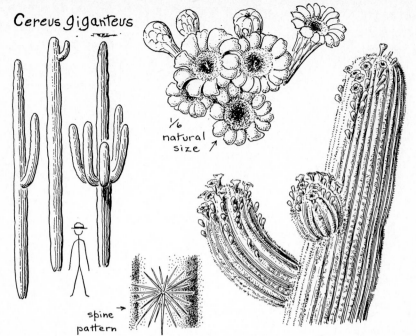

Cereus giganteus

1/6 natural size

spine pattern

Common Names: **SAGUARO (GIANT CACTUS).**
Arizona Desert: *(Cereus giganteus).* Waxy white. May-June.
Cactus family. Size: Up to 50 feet tall and 12 tons in weight.

Largest of the U. S. cactuses, this species occurs only in southern and western Arizona and adjoining northwestern Mexico and sparingly in extreme southeast California. It is an indicator of the Sonoran Desert.

This giant is such a spectacular example of desert vegetation that it is used as a trademark of the desert. It is the state flower of Arizona. Blossoms unfold at night, remaining open until late the following afternoon, attracting swarms of insects which in turn attract birds. Fruits mature in July, resembling small, egg-shaped cucumbers. When ripe, they burst open, revealing a scarlet lining and deep red pulp filled with tiny black seeds. Fruits are eagerly sought by birds and rodents.

Because of its enormous capacity for storing water in its spongy stem tissue, the saguaro (sah-WAR-oh) produces flowers and fruits even during droughts of long duration. When other foods failed, the Pima and Papago Indians could depend upon the saguaro harvest.

Saguaros are believed to live to a maximum age of 200 years, sometimes succumbing to a necrosis disease transmitted by the larvae of a small moth. Grazing cattle trample out the young plants and much of the desert occupied by saguaros is being placed under cultivation. Both Saguaro National Monument and Organ Pipe Cactus National Monument preserve and protect spectacular stands of these desert behemoths.

dry weather

rainy weather

skeleton

flesh

WHITE

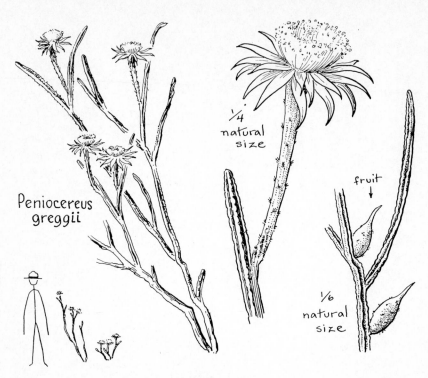

Peniocereus
greggii

¼ natural size

fruit

⅙ natural size

Common Names: **NIGHT-BLOOMING CEREUS, REINA-DE-LA-NOCHE, DEERHORN CACTUS.**

Arizona and Texas deserts. *(Peniocereus greggii).* White. June-July. Cactus family. Size: 2 to 5 feet tall.

One of the most delicately beautiful of the flowers for which the desert is famous, "queen of the night" is waxy-white with thread-like stamens that give it the appearance of wearing a halo. The night on which the cereus blooms is eagerly awaited by desert dwellers of long residence. All of the buds on a single plant, from two to six or seven in number, may open on the same night or may time their opening over a period of a week or more, usually in late June or early July, depending upon the season and other factors.

It is not unusual for nearly all of the plants in one locality to blossom on the same night. Buds unfold in the early evening, the flowers wilting permanently soon after sunrise the following morning. Fragrant, with a heavy, cloying perfume, they attract large numbers of night-flying insects.

The long, slender, fluted, lead-colored stems of the nightblooming cereus are inconspicuous and unattractive. They usually grow upward from beneath a creosotebush or other desert shrub, partially supported and almost entirely hidden by the larger plant.

The beet-like root, which serves as a moisture-storage organ, may weigh from 5 to 85 pounds and is reportedly eaten by desert Indians. Fruits are podlike, pointed at the ends, and the size of a large pickle. They turn dull red when mature.

WHITE

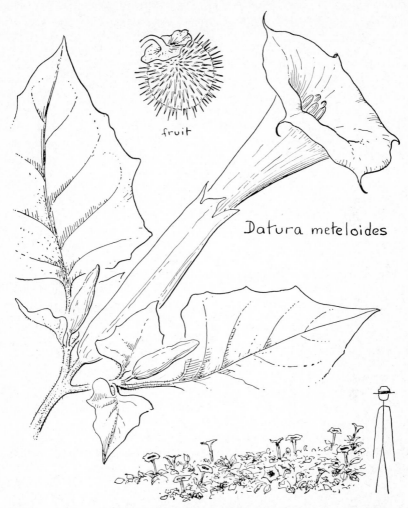

fruit

Datura meteloides

Common Names: **SACRED DATURA (WESTERN-JIMSON, THORN-APPLE, GIANT-JIMSON).**

Arizona, California, and Texas deserts. *(Datura meteloides).* White. May-October.

Potato family. Size: Up to 3 feet tall, and spreading over as much as 50 square feet of ground.

All portions of this coarse, vine-like herb are poisonous, and are used by some Indians as a narcotic to induce visions.

Seeds are sometimes administered to prevent miscarriage.

The plants with their large, gray-green leaves and showy, white, sometimes lavender-tinted flowers which open at night and close soon after contact by rays of the morning sun, are a common and arresting sight along roadsides and washes at elevations from 1,000 to 6,500 feet in Texas, New Mexico, Arizona, southern Utah, southern California, and Mexico.

WHITE

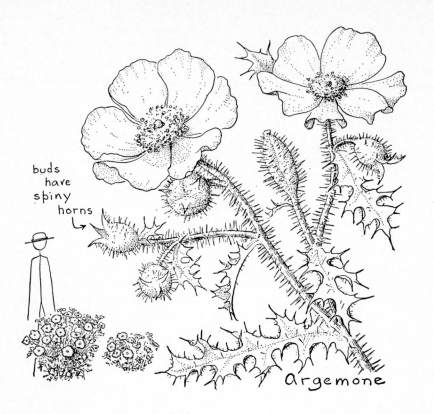

buds
have
spiny
horns

Argemone

Common Names: **CRESTED PRICKLEPOPPY (THISTLEPOPPY, CHICALOTE).**

Arizona, California, and Texas deserts. *(Argemone platyceras).* White. Blooms all year.

Poppy family. Size: Up to 30 inches in height.

One of the commonest and most noticeable perennials of the Southwest, the pricklepoppy ranges from South Dakota and Wyoming to Texas, Arizona, southern California, and northern Mexico. A coarse, prickly plant with large flowers and yellowish sap, it is easily recognized.

It is sometimes facetiously called "cowboys' fried egg."

Flowers are normally white with large, tissue-paper petals and yellow centers. In southern Arizona an occasional plant with pale yellow petals is found; and in Big Bend National Park, Texas, a form with rose-colored petals and a deep red center is occasionally encountered.

Plants are drought-resistant, unpalatable to livestock, and may be found in blossom during any month in the year, although much more prolific during the spring and summer. When abundant on cattle range, they are an indication of over-grazing. Seeds are reported to contain a narcotic more potent than opium.

WHITE   12

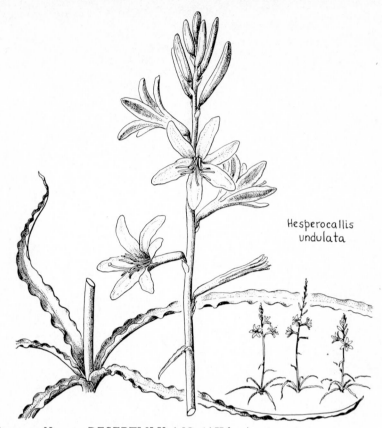

Hesperocallis
undulata

Common Names: **DESERTLILY AJO** (AH-hoe)
Arizona and California deserts. *(Hesperocallis undulata).* White. March-
April.
Lily family. Size: Narrow-leafed perennial, 6 inches to 2 feet.

One of the showiest and most famous of the desert wildflowers,
although limited in distribution to sandy areas below 2,000 feet
elevation, the desertlily greatly resembles the Easter-lily of green-
house habitat.

In some seasons the blossoms are abundant and their delicate
fragrance perfumes the surrounding atmosphere. During "off" seasons, visitors may scour the desert to find only a very few of the
fragile blossoms.

Named *"ajo"* by Spanish explorers because of the large, edible
bulb resembling garlic, this name has been given to a mountain
range, a broad valley, and a thriving town in southwestern Arizona,
where desertlily grows in profusion. Its range is limited to southwestern Arizona, southeastern California, and probably northern
Sonora.

Papago Indians eat the bulbs, which have an onion-like flavor.
Bulbs are difficult to obtain because they grow at a depth of 18
inches to two feet beneath the surface of the hard-packed desert
soil. Flowers remain open during the day, and propagation is
principally by seeds.

WHITE

Rafinesquia neomexicana

fruiting phase

Common Names: **DESERT CHICORY.**
Arizona and Texas deserts. *(Rafinesquia neomexicana).* Bright white.
  March-May.
California desert. *(Rafinesquia californica).* Dull white. April-May.
Sunflower family. Size: About a foot high.

In early springs that follow winters of more than average rainfall the desert-dandelion is one of the conspicuous annuals helping to carpet the deserts with a ground-cover of flowers.

Although much more delicate, longer stemmed, and less coarse and robust than the common dandelion, the flowers sufficiently resemble those of the better-known yellow dandelion to stimulate recognition.

Desert-dandelion is found below 4,000 feet in desert situations from western Texas to Lower California and northward to southern Utah.

WHITE   14

Nerium oleander

petal, to show odd crest

Common Name: **COMMON OLEANDER.**
Arizona, California, and Texas deserts. *(Nerium oleander).* White, yellow,
   or red. Spring and summer.
Dogbane family. Size: Robust, spreading shrub up to 20 feet high.

Well known and widely grown because of its large clusters of
red or white blossoms and glossy, evergreen leaves, the oleander is
one of the handsomest shrubs found under cultivation in towns and
cities of the desert. Requiring sub-tropical conditions, easily rooted
from cuttings, and rapid in growth, the oleander thrives in South
western desert areas if supplied with plenty of water. It is used
individually and as hedgerows in ornamental plantings.

Although blossoms may be present at almost any time of year.
the principal flowering season extends from early spring well
through the summer. Both the red-flowered and the white-flowered
plants are popular and may be grown separately or intermixed.
Recently a yellow-flowered form has come into use.

These handsome shrubs immediately attract the attention of
northerners visiting desert towns, and arouse their curiosity as to
their identity. Leaves are reported poisonous if eaten.

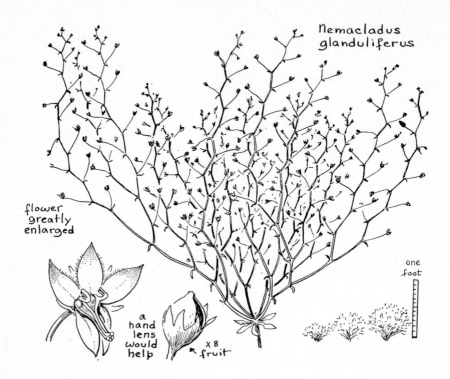

Nemacladus glanduliferus

flower greatly enlarged

one foot

a hand lens would help

x 8 fruit

Common Name: **THREADPLANT.**

Arizona desert. *(Nemacladus glanduliferus).* Purple-white. March-May
California desert. *(Nemacladus rigidus).* Purple-white. March-May.
Bellflower family. Size: 2 to 12 inches tall.

The tiny, slender-stemmed, profusely-branched threadplant is
so small that it is completely overlooked by the majority of visitors
to the Southwest, yet it is one of the most common and most attractive of desert flowers. Under a magnifying glass, the shape and
coloring of the minute, delicate flowers make them appear as
beautiful as orchids. The white flowers are touched with tints of red,
brown, yellow, or purple.

Plants are abundant below 1,800 feet elevation on dry, gravelly
or rocky soils, frequently along the shoulders of highways from
Nevada throughout western Arizona and southern California to
Lower California. Be on the lookout for this small but interesting
and beautiful plant.

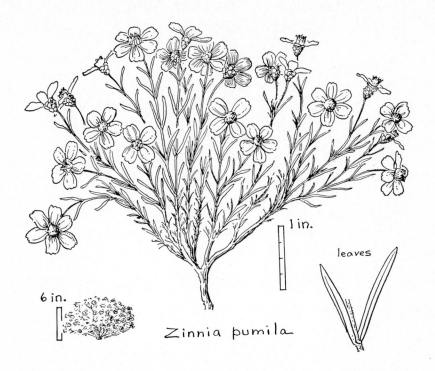

6 in.

1 in.

leaves

Zinnia pumila

Common Names: **ZINNIA (WILD ZINNIA).**

Arizona and Texas deserts. *(Zinnia pumila).* White. April-October.

Sunflower family. Size: Low, dense-growing perennials in rounded clumps, 4 to 8 inches high.

Closely related to the garden zinnia, which is a native of Mexico, desert zinnias are attractive herbs suitable for trial as ornamental border plantings.

Zinnia pumila prefers clayey soils and is found in dry mesas and slopes from Texas westward to southern Arizona and northern Mexico. It is often found blossoming in association with the paper-flower *(Psilostrophe cooperi)* which it superficially resembles. The white flowers of the wild zinnia turn cream with age.

Zinnia pumila may be easily recognized by the single heavy rib running the length of each narrow leaf. Blossoms may have four, five, or six petal-like rays.

WHITE

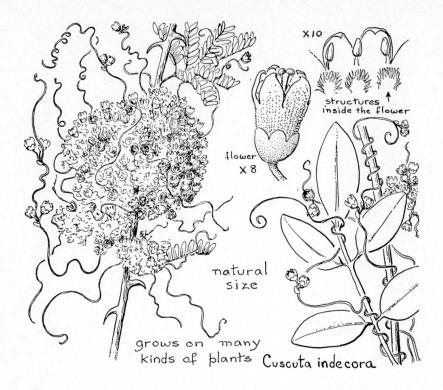

X10

structures inside the flower

flower X 8

natural size

grows on many kinds of plants — Cuscuta indecora

Common Name: **DODDER (BIGSEED ALFALFA DODDER).**
Arizona and Texas deserts. *(Cuscuta indecora)*. White. July-August.
California desert. *(Cuscuta denticulata)*. Pale yellow. July-August.
Convolvulus family. Size: Vine-like, covering host plant.

Rootless, leafless, and with pale yellow to brownish stems which twine in vine-like embrace about the host, the parasitic dodders are immediately noticeable because of their strange appearance.

Frequently the automobile traveler's attention is arrested by a pale yellowish blotch in the green of the roadside vegetation. Examination shows this to be caused by the matted yellowish stems and the white to pale yellow, fleshy blossoms. These flowers are attractive and often abundant enough to make a showy display.

Dodder is found widespread throughout the United States and is often a serious parasitic pest on crops of economic importance. Desert species are usually found infesting mesquite, goldenrod, aster, burrobush, seepwillow, and arrowweed. Although certain dodders show a preference in choosing hosts (C. *denticulata* common on creosotebush), most of them grow readily upon various plants.

WHITE

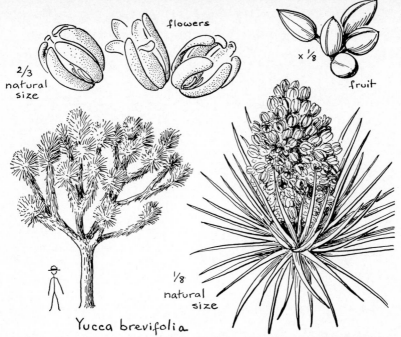

flowers

2/3 natural size

x 1/8 fruit

1/8 natural size

Yucca brevifolia

Common Names: **JOSHUA-TREE (TREE YUCCA, GIANT JOSHUA).**
California desert. *(Yucca brevifolia).* Green-white. February-April.
Lily family. Size: 15 to 35 feet high; spread of 20 feet.

Because the presence of this tree yucca marks, more effectively than any other plant, the limits and extent of the Mojave Desert, this species is worthy of special recognition. It holds, in the Mojave Desert, similar status to the saguaro in the Sonoran Desert. Strangely enough, in west-central Arizona the saguaro and Joshua-tree are found growing together and there the Sonoran and Mojave Deserts overlap.

And, just as in southern Arizona, an area has been set aside as Saguaro National Monument to preserve and protect that species, so in southern California we find the Joshua Tree National Monument.

The Joshua-tree is outstanding among the many species of yucca because of its short leaves growing in dense bunches or clusters, and because the plant has a definite trunk with numerous branches forming a crown. Great forests of these sturdy trees are found in parts of southern California, southern Nevada, southwestern Utah, and northwestern Arizona, where rainfall averages 8 to 10 inches per year.

Flowers of this yucca develop as tight clusters of greenish-white buds at the ends of the branches, but do not open wide as do the flowers of other yuccas. Joshua-trees do not bloom every year, the interval apparently being determined by rainfall and temperature. Birds, a small lizard, woodrats, and several species of insects are closely associated with this plant, making use of it for food, shelter, or nest-building materials. Indians use the smallest roots, which are red, for patterns in their baskets.

The name "Joshua-tree" was given by the Mormons because the tree seemed to be lifting its arms in supplication as did the Biblical Joshua.

WHITE

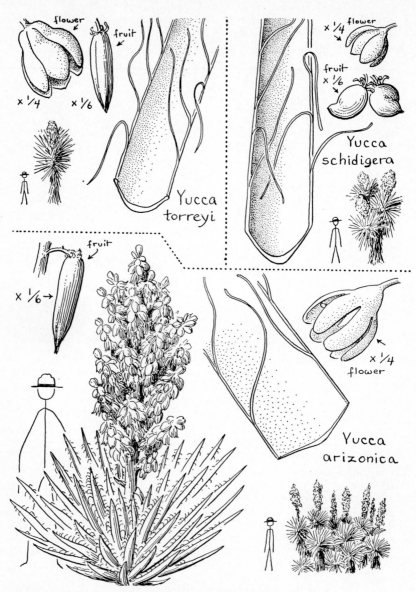

Common Names: **BROAD-LEAF YUCCAS, SPANISH BAYONET, AMOLE, DATIL, SOAPWEED.**

Arizona desert. *(Yucca arizonica).* Creamy. April-May.

California desert. *(Yucca schidigera).* White-purple. March-April.

Texas desert. *(Yucca torreyi).* Creamy. March-April.

Lily family. Size: Reaches height of 10 to 15 feet.

Although, in general, the broad-leafed yuccas do not reach tree size, the giant dagger *(Yucca carnerosana)* of Big Bend National

CREAM   20

Park reaches a height of 20 feet. In dense stands or "forests" these yuccas, with their huge clusters of creamy, wax-like, lightly scented, bell-shaped flowers produce a never-to-be-forgotten display in blooming season.

The yucca is the state flower of New Mexico.

Yuccas are often confused by newcomers to the desert with three other groups of plants: the *Agaves* ("centuryplant"), *Dasylirion* (sotol), and *Nolinas* ("beargrass, sacahuista").

The plate on the next page has been devoted to a comparison of the four groups, and by studying it carefully, the characteristics by which each may be identified can be determined.

Yucca leaf fibers have long been used by Indians for fabricating rope, matting, sandals, basketry, and coarse cloth. Indians also ate the buds, flowers, and emerging flower stalks. The large, pulpy fruits were eaten raw or roasted, and the seeds ground into meal.

Roots of the yuccas have saponifying properties and are still gathered by some tribes and used as soap, especially for washing the hair. Flowers are browsed by livestock. (See narrow-leaf yuccas and Joshua-tree yucca). *Yucca baccata,* a broad-leaf species found in the Southwest outside of the desert areas, is discussed in "Flowers of the Southwest Mesas."

Common Names: **NOLINA, SACAHUISTA (BEARGRASS, BASKETGRASS).**

Arizona desert. *(Nolina microcarpa)*. Tan-cream. May-June.

California desert. *(Nolina parryi)*. White-cream. May-June.

Texas desert. *(Nolina erumpens)*. White-cream. May-June.

Lily family. Size: Ragged clumps 4 to 8 feet in diameter and flower stalks up to 8 feet high.

The *Nolinas* are sometimes confused with sotol and the *Yuccas* and occasionally with the *Agaves*. However, the *Nolinas* resemble huge clumps of long-bladed grass, whereas sotol leaves are ribbon-like and yucca leaves taper to a sharp point. Flower stalks of the *Nolinas* are usually drooping and plume-like, and the numerous flowers are tiny. The many papery, dry-winged fruits often remain on the stalk until late autumn.

Nolinas do not grow on the flat mesas or sandy flats as do the yuccas, but are confined to exposed locations on rocky slopes above the 3,000-foot elevation. The Parry nolina of the California Desert is a larger and more spectacular plant than the species found

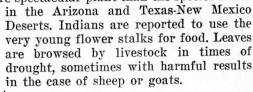

in the Arizona and Texas-New Mexico Deserts. Indians are reported to use the very young flower stalks for food. Leaves are browsed by livestock in times of drought, sometimes with harmful results in the case of sheep or goats.

Nolina
Parryi

CREAM

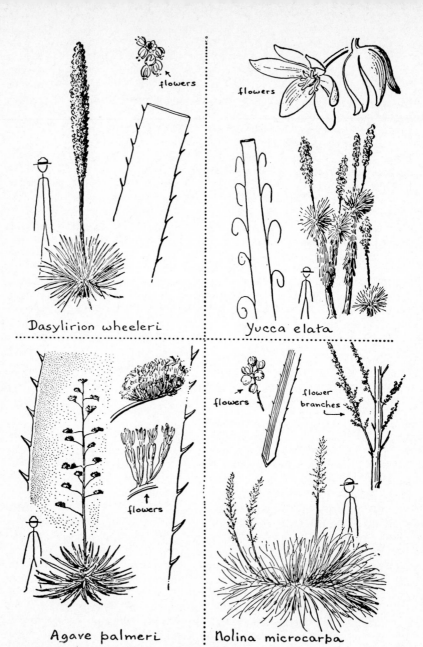

flowers

flowers

Dasylirion wheeleri

Yucca elata

flowers

flowers

flower branches

Agave palmeri

Nolina microcarpa

Common Names: **SOTOL (SPOONPLANT).**
Arizona desert. *(Dasylirion wheeleri).* Creamy. May-August.
Texas desert. *(Dasylirion leiophyllum).* Creamy. May-August.
Lily family. Size: Leaves 3 feet; flower stem 5 to 15 feet.

At first glance, this plant may readily be mistaken for a yucca, but its ribbon-like leaves (usually split at the tips instead of sharp-pointed) and tiny flowers instead of bell-like blossoms of the yucca, are distinguishing characteristics. The round heads of these plants grow close to the ground with the thick, woody stem beneath the

soil. Leaves, when stripped from the head come away with a broad, curved blade. When trimmed and polished, they are sold as curios called "desert spoons." In some desert areas near large cities, use of the plants for this purpose has endangered the species and aroused the ire of conservationists.

The cabbage-like base, after the leaves are removed, is split and fed to livestock as an emergency ration during periods of drought.

The rounded heads of these plants are high in sugar which is dissolved in the sap of the bud stalk. This sap, when gathered and fermented, produces a potent beverage called "sotol," which is the "bootleg" of northern Mexico.

Common Names: **NARROW-LEAF YUCCAS, SOAPTREE YUCCA, WHIPPLE YUCCA (PALMALLA, OUR-LORD'S-CANDLE, SPANISH-DAGGER, SOAPWEED, SPANISH-BAYONET).**

Arizona and Texas deserts. *(Yucca elata).* Creamy. May-July.
California desert. *(Yucca whipplei).* Creamy-white. May-June.
Lily family. Clumps 8-12 feet; *Yucca elata* sometimes to 20 feet.

The narrow-leaf yuccas are frequently confused with the *Agaves* ("centuryplant"), *Dasylirion* (sotol), and *Nolinas* ("beargrass") but may readily be recognized by the fibers protruding from the margins of the leaves. To permit comparison and bring out the differences so that the four groups may be recognized and confusion avoided, sketches of all four appear on the same plate (p. 22).

In many grassland areas of western Texas and southern New Mexico, *Yucca elata* dominates the landscape for miles. This species has been used as emergency rations for range stock during periods of drought, the chopped stems being mixed with concentrates such as cottonseed meal. A substitute for jute has been made from the leaf fibers. Indians eat the young flower stalks, which grow rapidly and are relatively tender.

In its relationship with a moth of the genus *Pronuba*, the yucca illustrates one of Nature's interesting partnerships. The moth, visiting yucca flowers at night, lays her eggs in the ovary of a flower, where the larvae will feed upon the developing seeds. To be sure that the seeds do develop, the moth must place pollen on the stigma of the flower. Dependent upon the moth for this vital act of pollination, the yucca repays its winged benefactor by sacrificing some of its developing seeds as food for the moth's larvae. Fruits of these narrow-leaved yuccas are dry capsules in contrast to the fleshy fruits of the broad-leaved ones.

Yucca whipplei, a much smaller plant than *yucca elata*, produces a stouter flower stalk with a great spreading plume of small, delicate flowers. These graceful plumes appear at night as if aglow with an inner light, hence the name "Our Lord's Candle." (See broadleaf yuccas [p. 20] and Joshua-tree yucca [p. 19].

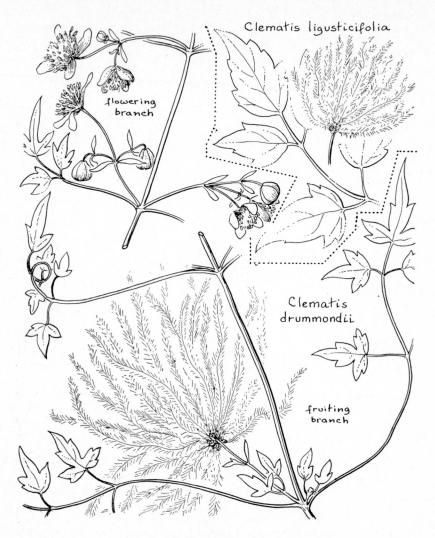

Clematis ligusticifolia

flowering branch

Clematis drummondii

fruiting branch

Common Names: **CLEMATIS, WESTERN VIRGINSBOWER (LEATHERFLOWER).**

Arizona and Texas deserts. *(Clematis drummondii)*. Cream. March-September.

California desert. *(Clematis ligusticifolia)*. Cream. Mid-September.

Crowfoot family. Size: Climbing, vine-like perennial with stems 6 to 8 feet long.

By no means limited to the desert, clematis is found throughout the Southwest. Several species are grown as ornamentals, foliage, flower-clusters, and the cotton-like masses of hairy fruits all being effecitve. Petals are absent or rudimentary, the sepals which furnish color to the blossoms being either creamy or purplish-brown. The name "leatherflower" has been applied to the latter group.

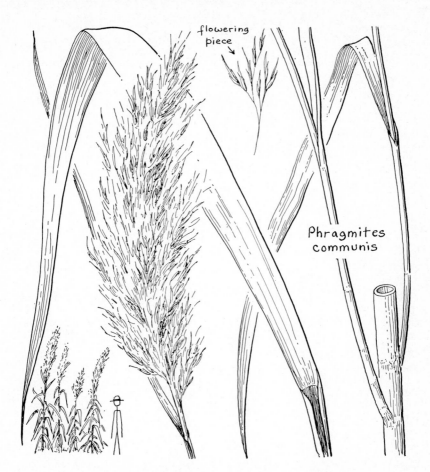

flowering
piece

Phragmites
communis

Common Names: **COMMON REED (CARRIZO, RIVERCANE, GIANTREED).**

Arizona, California, and Texas deserts. *(Phragmites communis).* Creamy. July-October.

Grass family. Size: 8 to 12 feet tall.

Among the largest of the grasses, the common reed and its close relative giantreed *(Arundo donax)* with their jointed stems resembling bamboo, are coarse perennials with broad, flat, grass-like leaves found in marshes and stock tanks, along irrigation canals and on river banks throughout the desert country of the Southwest. Common reed is found throughout the world where conditions are suitable. The flower stalks are long, tassel-like, and at the ends of the stems.

In Arizona and New Mexico, common reed is called *carrizo.* The hollow stems were used by the Indians for making arrow shafts, prayer sticks, pipestems, and loom rods. Mats, screens, nets, and cordage, as well as thatching, are made from the leaves. The plants are useful as windbreaks and in controlling soil erosion along streams.

CREAM

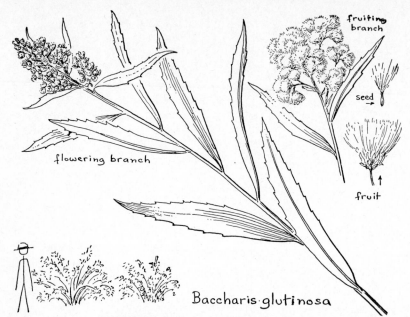

flowering branch

fruiting branch

seed →

fruit ↑

Baccharis glutinosa

Common Names: **SEEPWILLOW BACCHARIS, BROOM BACCHARIS (SEEPWILLOW, WATERMOTIE, WATERWALLY, WATERWILLOW, ROSIN-BUSH, HIERBA-DEL-PASMO).**

Arizona and Texas deserts. *(Baccharis glutinosa).* Creamy. March-December.

California desert. *(Baccharis sarothroides).* Yellow-white. September-February.

Sunflower family. Size: **Up to 7 feet tall.**

Genus *Baccharis* is composed, in the desert, of coarse shrubs with a number of common species. The flowers themselves are not beautiful, but the female plants with their flower heads that develop glaring-white pappus hairs, are spectacular and quite attractive.

Baccharis glutinosa is a common shrub along watercourses, often forming dense thickets. The straight stems are used in native houses as matting across ceiling timbers to support the mud roof. *Baccharis sarothroides* and several other species are often referred to as the desert brooms. They are common along desert washes and road-sides in sandy soil, their pale yellow, bristly flower heads, during the fall and winter months, appearing in sharp contrast to the vivid green branchlets and dark stems of the bushes. Among some Indians, these stems are chewed as a toothache remedy.

———

Common Names: **PLANTAIN, WOOLLY INDIANWHEAT (WOOLLY PLANTAIN)**

Arizona desert. *(Plantago purshii).* Buff. February-July.

California desert. *(Plantago insularis).* Straw. January-May.

Texas-New Mexico desert. *(Plantago argyrea).* Straw. June-August.

Plantain family. Size: A few inches to 2 feet tall.

Plantains are not noted for the beauty of their blossoms but the larger, coarser species are sufficiently noticeable to attract

CREAM

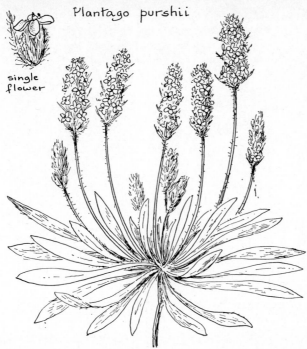

Plantago purshii

single
flower

attention, both in their blossoming and fruiting stages. The smaller winter annuals known as Indianwheat carpet the desert floor, in January and February, in some places, producing a straw-colored "pile" of tiny blossom spikes.

Common Names: **TESAJO, HOLYCROSS CHOLLA (TASAJILLO, CHRISTMAS CHOLLA, DIAMOND CACTUS, DARNING-NEEDLE CACTUS, PENCIL-JOINT CHOLLA).**

Arizona and Texas deserts. *(Opuntia leptocaulis)*. Green-yellow. May-June.
California desert. *(Opuntia ramosissima)*. Green-yellow. May-September.
Cactus family. Size: Much branched, shrubby, 2 to 4 feet tall.

Flowers of these small, slender-stemmed, shrubby chollas (CHOH-yahs) are small, sparse, and so inconspicuous as to be rarely noticed. However, the fruits, particularly those of *Opuntia leptocaulis*, are scarlet, egg-shaped, about 1 inch in length, and occur in such profusion that they immediately attract attention to the plants during the late fall and winter months, giving these plants the appropriate name of "Christmas" cholla.

A large cholla, *Opuntia bigelovii*, also has greenish to pale yellow flowers but inconspicuous fruits and short, heavy joints so densely covered with silvery spines as to give it the name teddybear cholla. Found in south central and southwestern Arizona and westward into southern California, southern Nevada, and south into Sonora and Lower California, the silver cholla is noticeable at any season. Propagation is chiefly by joints which drop from the plant and take root, the new plants forming dense thickets on desert hillsides. Because the joints are so easily detached, they actually seem to jump at a passerby, this characteristic giving the plant the name jumping cactus.

YELLOW

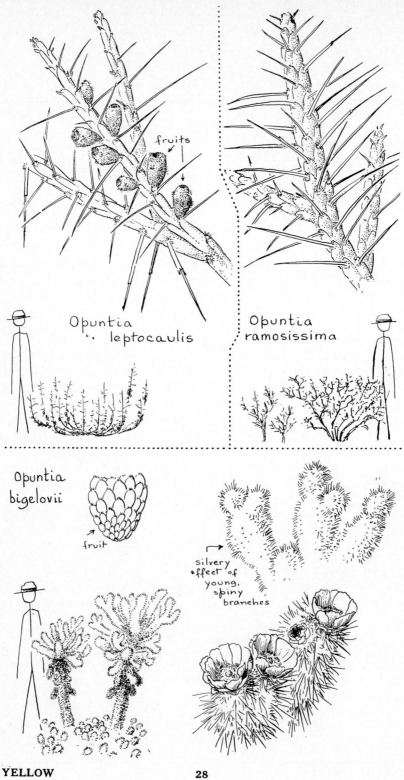

fruits

Opuntia
leptocaulis

Opuntia
ramosissima

Opuntia
bigelovii

fruit

silvery
effect of
young,
spiny
branches

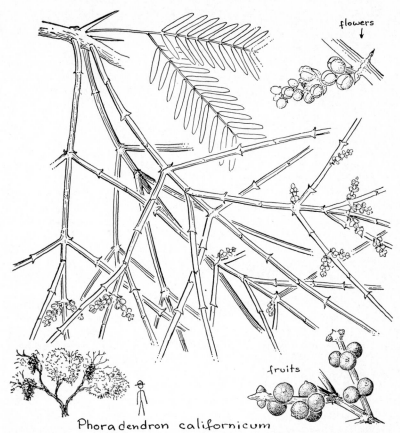

flowers

fruits

Phoradendron californicum

Common Names: **AMERICAN-MISTLETOE (DESERT MISTLETOE)**.
Arizona and California deserts. *(Phoradendron californicum)*. Yellow-green. March.
Texas-New Mexico deserts. *(Phoradendron cockerellii)*. Yellow-green. Spring.
Mistletoe family. Size: Pendant, vine-like strands several feet long.

Because they form conspicuous, dense, shapeless masses in mesquite, ironwood, acacia, cottonwood, or other trees (depending upon the species of mistletoe), these parasitic plants attract the attention and arouse the curiosity of persons unfamiliar with the desert. *Phoradendron macrophyllum*, which parasitizes cottonwood trees, is widespread throughout the Southwest, and because of its large gray-green leaves and glistening white berries is much in demand as a Christmas green. The mistletoe is the state flower of Oklahoma.

The species of mistletoe that parasitize such trees as ironwood, mesquite, and catclaw have small, scale-like, tawny-brown leaves and stems. The tiny, yellow-green flowers which appear in spring are fragrant, and secrete nectar which attracts honeybees and other insects. The handsome, coral-pink berries are a major food, during the winter months, for phainopeplas and other birds. The Arizona verdin often builds its nest in the protected center of a clump of mistletoe. Birds are believed to be instrumental in spreading this parasite from tree to tree.

American-mistletoe saps the energy of the host tree and, where abundant, may cause considerable damage, killing branches and sometimes the entire tree. Papago Indians dry the berries in the sun and store them for winter food.

Common Names: **TREE TOBACCO, DESERT TOBACCO.**
Arizona and Texas deserts. *(Nicotiana glauca).* Pale yellow. All year.
California desert. *(Nicotiana trigonophylla).* Green-yellow. All year.
Potato family. Size: Tree tobacco *(Nicotiana glauca)* up to 12 feet. Desert
   tobacco, 1 to 3 feet high.

Several species of wild tobacco are found in the desert. Of these, tree tobacco is conspicuous because of its rank growth, its large leaves, and the spectacular clusters of tubular, yellow flowers. In addition to nicotine, tree tobacco contains an alkaloid, anabasine. This conspicuous plant occurs in moist locations below 3,000 feet elevation and bears flowers throughout the entire year. Although now thoroughly naturalized in the Southwest, it is a native of South America.

Desert tobacco, sometimes perennial in southwestern Arizona, is a dark green herb common and widespread throughout the desert areas of the Southwest. It is not nearly as noticeable as its larger relative although it, too, blossoms the year around. Flowers are a pale yellow, almost greenish-white. It provides dense ground cover in rocky canyons and along desert washes.

Leaves, which are somewhat bad smelling, were smoked (and still are during ceremonials) by the Yuma and Havasupai Indians. who are reported to have cleared land, burned the brush, and scattered the seeds of desert tobacco in an effort to promote the growth of strong plants with many large leaves.

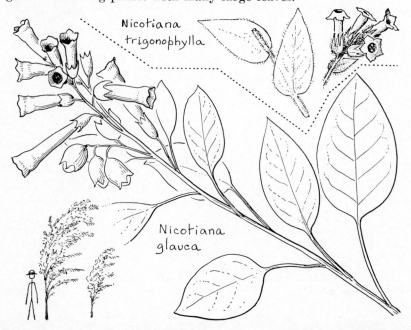

Nicotiana
trigonophylla

Nicotiana
glauca

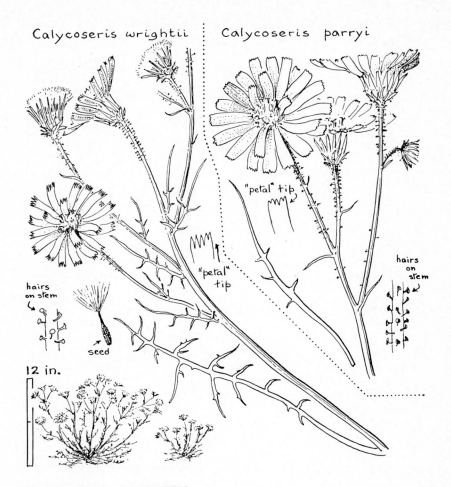

Calycoseris wrightii   Calycoseris parryi

"petal" tip

"petal" tip

hairs on stem

hairs on stem

seed

12 in.

Common Name: **TACKSTEM**
Arizona desert. *(Calycoseris wrightii)*. White. March-May.
California desert. *(Calycoseris parryi)*. Yellow. March-April.
Sunflower family. Size: 4 inches to a foot tall.

One of the handsomest of desert spring annuals, *Calycoseris* is common on plains, mesas, and rocky slopes at elevations between 1,200 and 4,000 feet from western Texas to southern Utah, southern California, and South into Mexico.

The name tackstem comes from the presence of numerous tack-shaped glands which protrude from the stems.

Taking advantage of the cool, moist weather of winter, the tackstems produce their beautiful rose, white, or yellow blossoms in early spring, and mature their seeds before the advent of hot, dry weather.

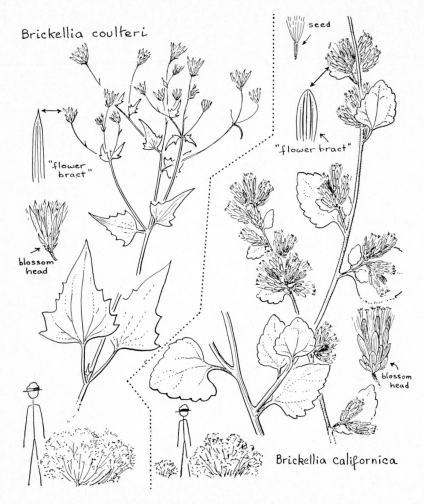

Brickellia coulteri

seed

"flower bract"

"flower bract"

blossom head

blossom head

Brickellia californica

Common Names: **BRICKELLIA (BRICKELLBUSH, DESERT BRICKELLIA, PACHABA).**

Arizona desert. *(Brickellia coulteri).* Yellow-white. September.

California desert. *(Brickellia desertorum).* Pale yellow. Midsummer.

Texas-New Mexico deserts. *(Brickellia californica).* Yellow-white. July-October.

Sunflower family. Size: Small, much-branched perennial shrub, up to 3 feet in height.

Intricately branched and brittle-stemmed, this shrub with blossom heads holding from 8 to 18 yellowish flowers is common throughout the Southwest from western Texas and Colorado to Nevada, Sonora and Lower California.

It grows among rocks and in rocky locations throughout much of the desert country from 3,000 up to 7,000 feet.

YELLOW   32

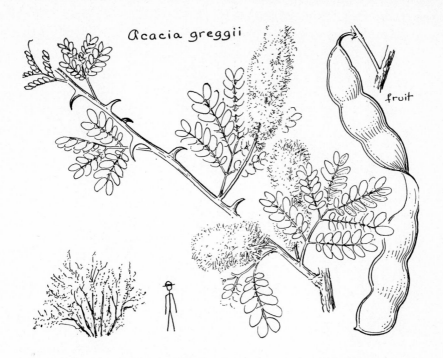

Acacia greggii

fruit

Common Names: CATCLAW ACACIA (CAT'SCLAW, TEARBLANKET, DEVIL'SCLAW).

Arizona, California, and Texas deserts. *(Acacia greggii).* Pale yellow. April-October.

Pea family. Size: Up to 20 feet tall.

The numerous thorns, short and curved like a cat's claw, serve readily to identify this common, often abundant, shrub or small tree.

There are several species, some with large, bright-yellow flowers, but *Acacia greggii* is the most common and occurs throughout all of the deserts of the Southwest, at elevations below 4,000 feet, often forming thickets along streams and washes.

Flowers, like pale yellow, fuzzy caterpillars, are one of the important sources of nectar for honeybees, the trees being alive with insects during the period of heaviest blooming in April and May.

In mid-August, the light green fruit pods begin to turn reddish and, if abundant, make a colorful display.

Seeds of catclaw acacia were at one time widely used as food by the Indians of Arizona and Mexican tribes. They were ground into meal and eaten as mush or cakes.

Catclaw acacia is one of the most heartily disliked shrubs in the Southwest, especially by riders and hikers, because of the strong thorns which tear clothing and lacerate the flesh.

YELLOW

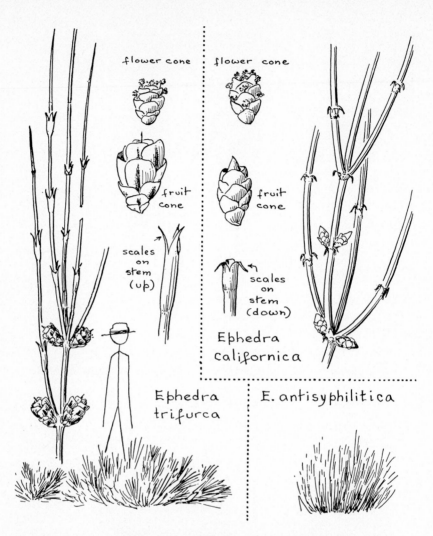

flower cone | flower cone

fruit cone

fruit cone

scales on stem (up)

scales on stem (down)

Ephedra californica

Ephedra trifurca

E. antisyphilitica

Common Names: **LONGLEAF EPHEDRA, CALIFORNIA EPHEDRA, (JOINTFIR, MORMON-TEA, POPOTILLA, TEPOSOTE, CANATILLA).**

Arizona and Texas deserts. (*Ephedra trifurca*). Pale yellow. Spring.

California desert. (*Ephedra californica*). Pale yellow. Spring.

Jointfir family. Size: Tough, stringy perennials, from 2 to 10 feet tall and sometimes 5 or 6 feet in diameter.

Apparently leafless, these common Southwestern shrubs do have leaves, although they are reduced to tiny scales. The harsh, stringy stems are green to yellow-green and, when dried, were used with the flowers in making a palatable brew, particularly by the Utah pioneers; hence the name Mormon-tea. The beverage was also popular with Indians and settlers in treating syphilis and other afflictions, as it contains tannin and certain alkaloids. Flowers are small, pale yellow, and appear in the spring at which time the plants are quite noticeable, and attract large numbers of insects.

———

Common Names: **CAMPHOR-WEED, TELEGRAPH PLANT**
Arizona and Texas deserts. *(Heterotheca subaxillaris).* **Pale yellow.** March-November.
Sunflower family. Size: Grows 2 to 6 feet tall.

The flowers are not particularly attractive, but become conspicuous as the seed-heads develop, because of the white, densely-haired tufts. Stems are tall and straight "like telegraph poles," and the crushed leaves give off a strong, somewhat camphor-like odor.

Although the plant occurs from the east coast across the southern portion of the United States, it is found in the desert at elevations between 1,000 and 5,000 feet.

Camphor-weed is a tall, coarse, robust, straight-stemmed plant which is abundant and conspicuous along roads and ditchbanks, and in the open desert following winters of heavy precipitation.

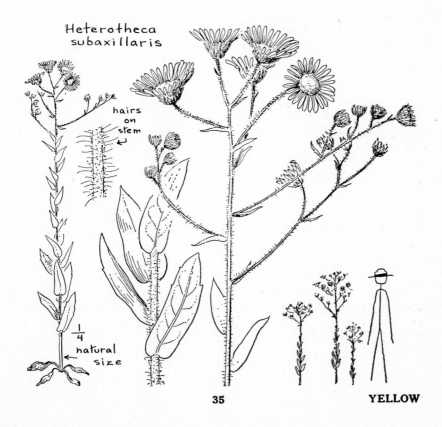

Heterotheca subaxillaris

hairs on stem

1/4 natural size

YELLOW

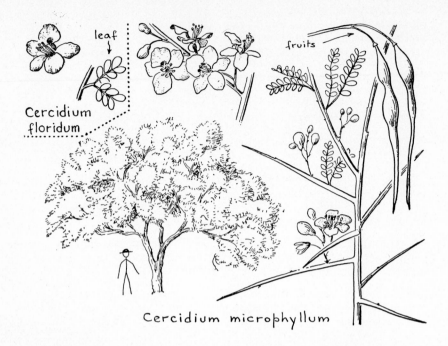

leaf

fruits

Cercidium floridum

Cercidium microphyllum

Common Names: PALOVERDE, LITTLELEAF PALOVERDE, BLUE PALOVERDE, HILLSIDE PALOVERDE

Arizona desert. *(Cercidium microphyllum)*. Pale yellow. April-May.
Arizona desert. *(Cercidium floridum)*. Bright yellow. April-May.
Pea family. Size: Green-barked tree up to 25 feet high.

Arizona paloverdes (meaning green stick) are large shrubs or small trees abundant along washes in the hotter, drier portions of the Sonoran Desert. When in blossom in the springtime, they appear as masses of pale yellow or golden bloom, and are a glorious sight, both as individual trees and massed as borders along the courses of washes which they mark with a line of color winding across the desert floor. During the dry season, they are without leaves, but are readily recognized by the bark, yellowish green in the case of *Cercidium microphyllum;* blue green in *Cercidium floridum.*

After the petals form, seeds develop in bean-like pods which are not relished by livestock, but are eaten during periods of drought and when other forage is scarce. Indians ground the seeds into meal.

When the trees are in blossom they attract myriads of insects, some of which, including honeybees, seek the nectar. Wood is soft and the branches are brittle and easily broken. It is unsuited for fuel, as it burns rapidly, leaves no coals, and gives off an unpleasant odor.

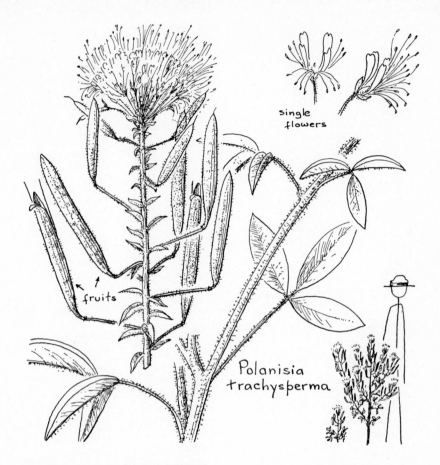

single flowers

fruits

Polanisia trachysperma

Common Name: **CLAMMYWEED**
Arizona desert. *(Polanisia trachysperma).* Pale yellow. June-September.
Texas desert. *(Polanisia uniglandulosa).* Pale yellow. June-September.
Caper family. Size: 1 to 3 feet tall.

Clammyweed is not limited in its range to desert areas, but is found as far north as Saskatchewan and British Columbia. However, it is also a common annual in Texas, New Mexico, and Arizona at elevations between 1,200 and 6,000 feet, and is usually found in abundance in the sandy channels of dry stream beds.

It somewhat resembles both yellow spiderflower *(Cleome lutea)* and jackass-clover *(Wislizenia refracta).*

YELLOW

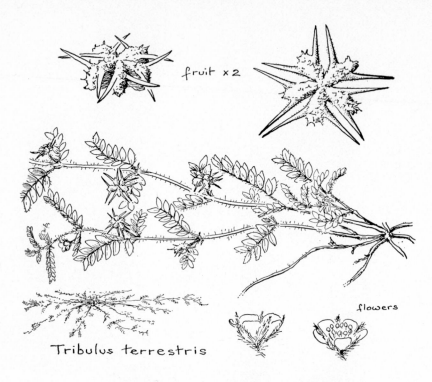

fruit x2

flowers

Tribulus terrestris

Common Names: PUNCTUREVINE (BURNUT, BULLHEAD, TORRITO).

Arizona, California, and Texas deserts. *(Tribulus terrestris)*. Pale yellow. Summer.

Caltrop family. Size: Prostrate, stems 2 to 6 feet long.

A troublesome annual vine-like weed naturalized from southern Europe, the puncturevine has established itself throughout the Southwest below 7,000 feet. Although fairly readily controlled by cultivation, the plant spreads rapidly in sandy, dry wastelands, often taking over vacant lots in towns, and areas in the desert where it finds sufficient moisture.

The fruits, which are produced in quantities, are armed with strong spurs which become embedded in the feet and fur of animals and in automobile tires. Fruits are also carried by irrigation or flood waters. Although the spurs are too short to puncture automobile tires, they make bicycles almost useless in some localities, and are an aggravation to children who go barefoot—and to dogs.

Flowers and fruits in various stages of maturity may be found on this fast-growing plant at almost any time during the summer months. Botanically, puncturevine is closely related to the creosotebush and also to the Arizona-poppy.

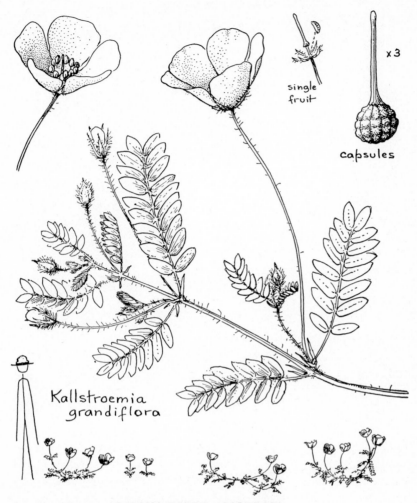

single fruit

x3

capsules

Kallstroemia
grandiflora

Common Names: **ARIZONA-POPPY (CALTROP).**

Arizona and Texas deserts. *(Kallstroemia grandiflora)*. Bright yellow. February-September.

Caltrop family. Size: 1 to 2 feet tall.

Although superficially resembling in size, shape and color the blossoms of the goldpoppy, the blossoms of the large-flowered caltrop have five petals instead of four, and the plant is a close relative of the puncturevine and the creosotebush. One of the most attractive of the desert's summer annuals, Arizona-poppy is found at elevations below 5,000 feet in the drylands of Texas, New Mexico, Arizona, and northern Mexico.

Large-flowered caltrop may be distinguished from goldpoppy by (1) sprawling open habit of growth, (2) compound leaves, (3) season of blossoming, and (4) the fact that the plants grow singly rather than in masses.

flower

fruit

×4

leaves

anthers and scales

Larrea tridentata

Common Names: CREOSOTEBUSH (GREASEWOOD, HEDIONDILLA).
Arizona, California, and Texas deserts. *(Larrea tridentata)*. Yellow. Spring.
Caltrop family. Size: Shrub, 2 to 8 feet high.

No one could justifiably question the statement that creosotebush is the most successful, widespread, and readily recognized
desert plant of the hot, arid regions of North America. It often
occurs over wide areas in such pure stands as to constitute true
Larrea plains. Its common companion is the grayish burrobush or
bursage.

Following winter rains, the creosotebush may put out a few
yellow blossoms in January, but usually burst into full flower in
April or May, to be followed in a short time by the equally spectacular fuzzy white seed balls making the bushes appear to be
covered with a light frosting of snow. After a rain, the plants give
off a musty, resinous odor which is the basis of the Mexican name
hediondilla (freely translated, "little stinker"). Lac occurs as a
resinous incrustation on the branches, and was used by the Indians
for mending pottery, making mosaics, and for fixing arrow points.

Leaves of the creosotebush are covered with a "varnish" which often glistens in the sunlight, and helps reduce evaporative moisture loss, thereby enabling the plant to resist the desiccating effect of hot, dry winds.

———

Common Name: **JACKASS-CLOVER**

Arizona desert. *(Wislizenia refracta).* Yellow. May-September.

Caper family. Size: Up to 4 feet in height.

Conspicuous in late summer along roadsides and dry streambeds, the large number of yellow flowers and the widespread presence of these much branched, annual plants justify the inclusion of jackass-clover in this booklet as one of the common flowers of the desert.

The plant ranges across the Southwest from western Texas to southern California at elevations between 1,000 and 6,500 feet. The flowers themselves are small, although the flower heads are quite conspicuous.

Since the leaves somewhat resemble the tri-foliate leaves of clover, the plant is commonly called jackass-clover. It is usually found in sandy locations.

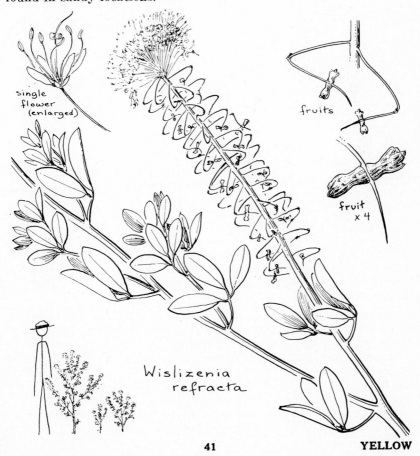

single flower (enlarged)

fruits

fruit x 4

Wislizenia refracta

YELLOW

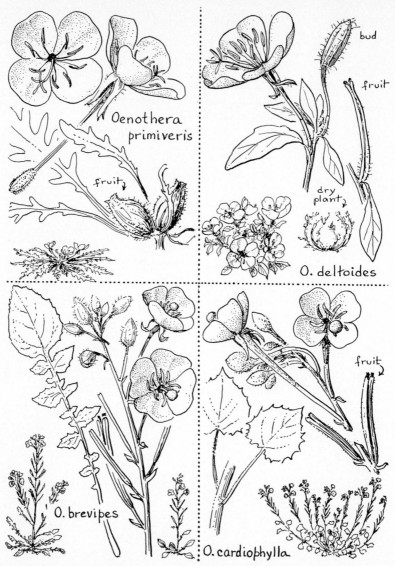

Oenothera primiveris

O. deltoides

O. brevipes

O. cardiophylla

Common Names: EVENING-PRIMROSE, SUNDROP
Arizona and California deserts. *(Oenothera brevipes)*. Yellow. March-May.
Texas-New Mexico deserts. *(Oenothera primiveris)*. Yellow. March-May.
Evening-primrose family. Size: Usually low, but some up to 5 feet.

Among the commonest but most beautiful and delicate of desert flowering plants are the evening-primroses. Flowers are usually large, with the four petals either white or yellow, turning to red or pink with age. Many species are low-growing herbs with large, delicate petals; while others may be shrub-like, sometimes attaining a height of 5 feet. As the name implies, the flowers open in the evening and wilt soon after sunrise.

In the low, warmer sections of the desert, plants in blossom may be found as early as February.

Common Names: **BARBERRY, ALGERITA, (MAHONIA,
HOLLYGRAPE).**
Arizona desert. *(Berberis haematocarpa).* Yellow. February-April.
California desert. *(Berberis fremontii).* Yellow. May-July.
Texas desert. *(Berberis trifoliolata).* Yellow. Spring.
Barberry family. Size: Shrubs, 3 to 8 feet.

The pendant clusters of golden blossoms are particularly noticeable because of their delightful fragrance, and the small purple
berries are juicy and of pleasant flavor. They make excellent jelly
and are readily eaten by birds and some of the small mammals.
Due to the holly-like leaves and the fragrant blossoms and fruits,
the plants would make attractive ornaments for landscape and
decorative plantings were it not for the fact that they are secondary
hosts for the black stem rust of the cereals, hence cannot be used
in communities where grains are grown. Indians use the root as a
tonic, and obtain from it a brilliant yellow dye.

Some botanists prefer to use the generic name *Mahonia* or *Odo-
stemon* for this group of plants.

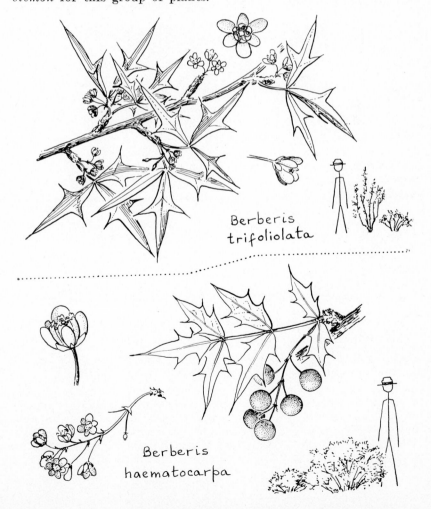

Berberis
trifoliolata

Berberis
haematocarpa

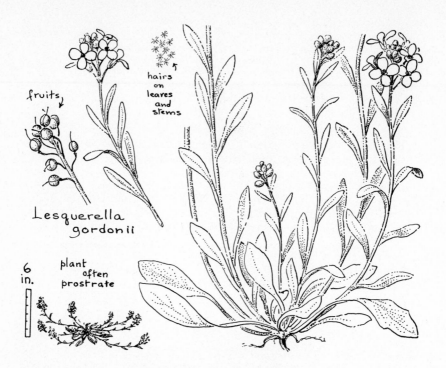

fruits

hairs on leaves and stems

Lesquerella gordonii

6 in.

plant often prostrate

Common Names: **BLADDERPOD (BEAD-POD).**
Arizona and Texas deserts. *(Lesquerella gordonii).* Yellow. February-May.
California desert. *(Lesquerella palmeri).* Yellow. March-May.
Mustard family. Size: 6 to 8 inches high.

Extensive sections of the desert are gilded in springtime with this low-growing annual herb which is one of the earliest of the desert flowers.

Following moist winters, it covers dry mesas and plains below 4,000 feet from Oklahoma west to Utah, and southward into northern Mexico. After the seed pods have matured, the plant is reported to furnish valuable forage for range stock.

———

Common Names: **BUFFALO-GOURD (COYOTE-MELON,**
CALABAZILLA, CHILI COYOTE).
Arizona desert. *(Cucurbita digitata).* Yellow. June-October.
California desert. *(Cucurbita palmata).* Yellow. July-September.
Texas desert. *(Cucurbita foetidissima).* Yellow. May-August.
Gourd family. Size: Trailing perennial with stems 4 to 15 feet long.

Gourds are conspicuous, trailing, rank-growing plants common along roadsides and in the open desert. Leaves are grayish-green, and blossoms yellow and trumpet-shaped. The striped fruits are about the size and shape of a tennis ball, although some are egg-shaped.

YELLOW

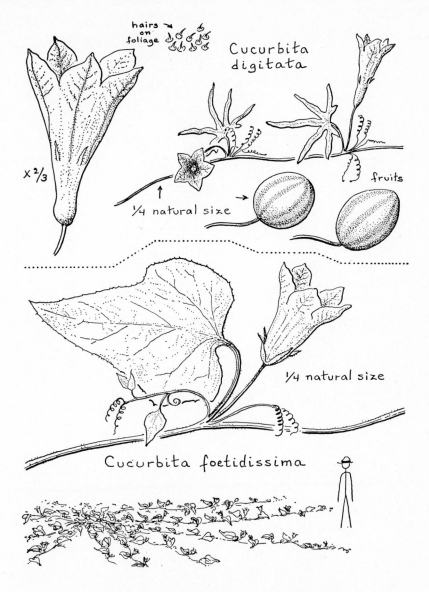

hairs on foliage

Cucurbita digitata

X ⅔

¼ natural size

fruits

¼ natural size

Cucurbita foetidissima

The fruits, which are very conspicuous after the vines and leaves have been winter-killed, are sometimes collected, painted in gay colors, and used as ornaments about the house.

Although Indians considered the fruits as inferior and suitable only for coyotes, they ate them cooked or dried, and made the seeds into a mush. Pioneers used the crushed roots of these plants as a cleansing agent in washing clothes.

YELLOW

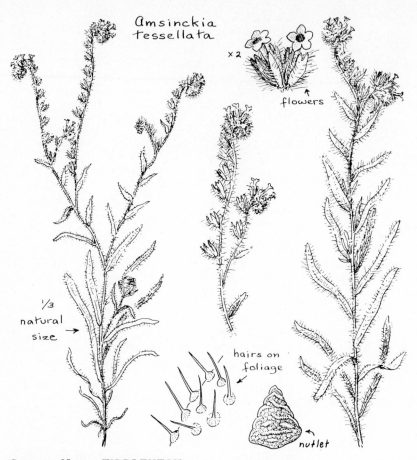

Amsinckia tessellata

×2

flowers

⅓ natural size →

hairs on foliage

nutlet

Common Name: **FIDDLENECK**
Arizona desert. *(Amsinckia intermedia)*. Yellow. Spring.
California desert. *(Amsinckia tessellata)*. Yellow. Spring.
Borage family. Size: Bristly erect herbs, 8 to 18 inches.

An annual of the creosotebush belt, and very abundant on gravelly or sandy soils in dry, open places. Fiddleneck is found from western New Mexico to California and north to eastern Washington.

Amsinckia tessellata occurs also in Chile and Argentina. Plants are reported to make good spring forage where they grow in heavy stands, but indications have been found that cirrhosis of the liver may result in cattle, sheep and horses that eat the nutlets.

Following moist winters, fiddleneck is often so abundant as to form vast fields of yellow or orange-yellow blossoms. especially on the Mojave Desert in southern California.

The curling habit of the opening flower heads somewhat resembles the neck of a violin, hence the name.

YELLOW   46

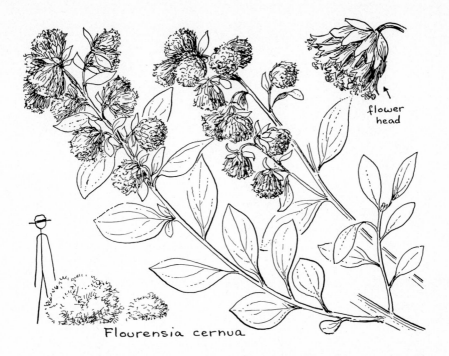

Flourensia cernua

Common Names: **AMERICAN TARBUSH (VARNISHBUSH).**
Arizona and Texas deserts. *(Flourensia cernua)*. Yellow. July-December.
Sunflower family. Size: A small shrub 3 feet, occasionally 6 or 7 feet high.

These resinous, much-branched, perennial shrubs are found on plains and mesas at elevations around 4,000 feet from western Texas to eastern Arizona and south into Mexico. The yellow, nodding flower heads are small, and the leaves have a hop-like odor and a bitter flavor unpalatable to cattle.

In northern Mexico the leaves and dried flower heads are sold in the drug markets under the name of *hojase*, recommended, in the form of a brew, as a remedy for indigestion.

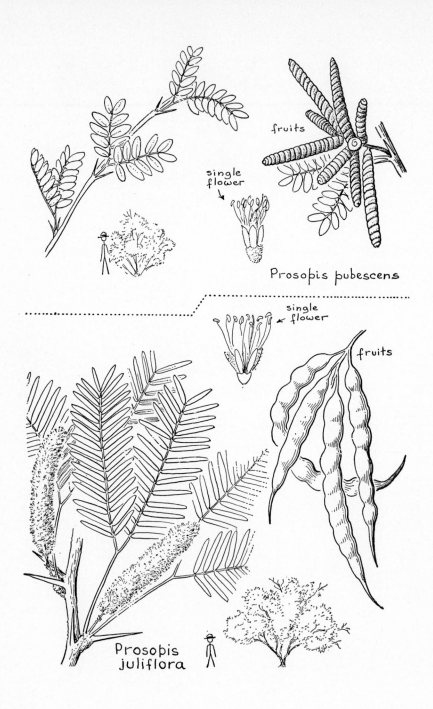

fruits

single flower

Prosopis pubescens

single flower

fruits

Prosopis juliflora

Common Names: **MESQUITE, HONEY MESQUITE**
Arizona, California and Texas deserts. *(Prosopis juliflora).* Yellow.
   April-June.
Pea family. Size: Tree 15 to 25, rarely 30 feet high.

Mesquite (mess-KEET) is one of the commonest and most wide-spread of desert trees, often growing in extensive thickets. It occurs at elevations below 5,000 feet, usually along streams, desert washes, or in locations where the water table is relatively high, from Kansas to California and south into Mexico. Roots are reported to penetrate to a depth of 60 feet with more wood below ground than above. In some parts of the desert, blowing sand settles around mesquite clumps forming hummocks through which rodents tunnel.

The numerous branches are armed with sturdy, straight thorns. In the spring when covered with bright green leaves and laden with catkin-like clusters of greenish-yellow flowers, mesquite is a particularly handsome shrub or tree. Blossoms are fragrant and attract myriads of insects, including honeybees.

During pioneer days, mesquite wood was of the utmost importance to settlers as fuel, and was also used extensively in building corrals and in making furniture and utensils. With the exception of ironwood (see page 81), mesquite is the best firewood to be found in the desert, giving off a characteristic aroma and forming a long-lived bed of coals.

Fruits of the mesquite, which resemble string beans, ripen in autumn and are eaten by domestic livestock and other animals. They are rich in sugar and still form a staple food among natives. Indians made wide use of mesquite, the fruits often carrying them over periods when their crops failed. *Pinole,* a meal made by grinding the long, sweet pods, was served in many ways. When fermented, it formed a favorite intoxicating drink of the Pimas. The gum, which exudes through the bark, was eaten as candy, and was used as a pottery-mending cement, and as a black dye.

Common Names: **SCREWBEAN MESQUITE (FREMONT SCREWBEAN,**
         SCREWPOD MESQUITE, TORNILLO).
Arizona, Texas, and California deserts. *(Prosopis pubescens).* Yellow. May-
   June.
Pea family. Size: Shrub, or tree up to 20 feet.

Although the screwbean, so called because of the tight spiral curl formed by the seed pod, is not as common as honey mesquite, it is nearly as widespread, being found below 4,000 feet from western Texas to southern Nevada, and southern California to northern Mexico. The majority of the trees are small and shrubby.

Fruits, in common with those of honey mesquite, are used by Indians and livestock for food. Bark from the roots was used by the Pima Indians to treat wounds. Where abundant, the wood is used for fence posts, tool handles, and fuel. Birds, particularly the crissal thrasher, make use of the shreddy bark for nest-lining material.

Where screwbean mesquite and honey mesquite grow together, they may be distinguished in the winter when trees are leafless and fruits have fallen or been removed by animals, by the gray-barked twigs of the former, those of the latter being brownish red.

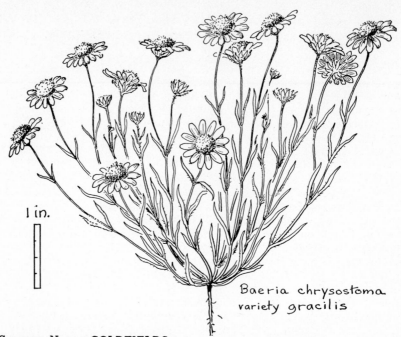

Baeria chrysostoma
variety gracilis

Common Name: **GOLDFIELDS**
Arizona desert. *(Baeria chrysostoma)*. Yellow. March-May.
Sunflower family. Size: Low growing, usually under 6 inches.

After winters of particularly heavy precipitation, these small close-growing annuals with their sunflower-like blossoms cover large patches of desert with a carpet of gold. Individual flowers are so small and so inconspicuous among larger plants that they are easily passed unnoticed, but millions of the plants all in blossom at the same time make a spectacular display that attracts visitors from considerable distances.

They occur in Arizona below 3,600 feet, westward to California, Lower California, and north to Oregon. A plant of winter and early springtime, goldfields takes advantage of winter moisture and cool spring weather to produce its flowers and mature its seeds. Thus it escapes the heat and drought of the desert by lying dormant in the seed stage until the moisture and cool temperatures of the following winter awaken it.

In common with goldpoppy and other annuals that mature their seeds before the summer heat descends upon the desert, goldfields cannot correctly be called a "desert plant." Actually these are plants of cooler climes which have found winter conditions in the desert ideal for their needs and have established themselves.

These plants demonstrate effectively one method, that of escaping the heat and drought, by which plants have adapted themselves to survival in the desert. Like the winter tourist, they take advantage of ideal climatic conditions of winter and spring. Since, unlike the winter tourist, they cannot return north for the summer, they take the next best course and pass through the hot, dry period in the dormancy of the seed phase of their life cycles.

Common Names: **ENCELIOPSIS (SUNRAY).**

Arizona desert. *(Enceliopsis argophylla)*. Bright yellow. April-June.

California desert. *(Enceliopsis covillei)*. Lemon-yellow. April-June.

Sunflower family. Size: Perennial, 1 to 2½ feet tall.

The large, solitary, coarse flower heads with their yellow petal-like rays make the enceliopsis among the most impressive composites of the desert.

Flowers rise on stout stems above a luxuriant growth of leaves that make the plants appear almost egotistical in their elegant arrogance.

They are at their best in sandy washes and on dry slopes at elevations between 1,000 and 3,500 feet, often where other plants seem too hard pressed eking out an existence to produce the garish foliage and bloom achieved by the "sunray."

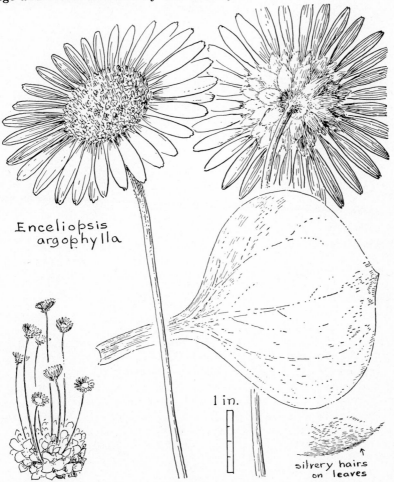

Enceliopsis
argophylla

1 in.

silvery hairs
on leaves

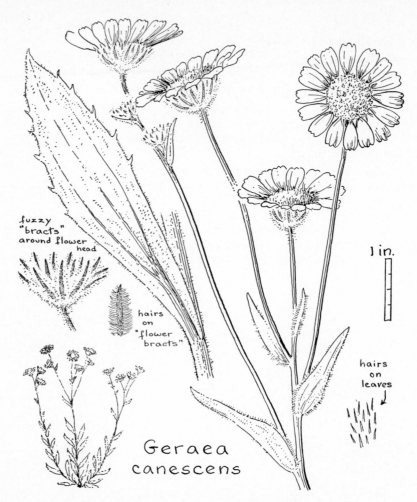

fuzzy "bracts" around flower head

hairs on "flower bracts"

1 in.

hairs on leaves

Geraea canescens

Common Names: **DESERTGOLD (DESERT-SUNSHINE, DESERT-SUNFLOWER, HAIRY-HEADED SUNFLOWER).**

Arizona, California, and Texas deserts. *(Geraea canescens).* Yellow. January-June.

Sunflower family. Size: An annual, 6 inches to 2 feet tall.

One of the showiest of the sunflowers. Desertgold often forms sweet-scented gardens of luxuriant bloom along roadsides and in sandy basins early in the spring.

Its seeds form a dependable source of food for small rodents, especially pocket mice, which store them in quantities. Wild bees and hummingbird moths are attracted to the fragrant flowers.

This species is common in areas of sandy soil below 3,000 feet in elevation from Utah and southeastern Colorado to southern Arizona and Sonora, Mexico. It is one of the showy roadside **flowers of** Organ Pipe Cactus National Monument.

YELLOW   52

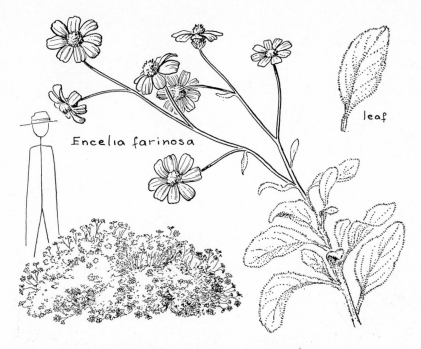

Encelia farinosa

leaf

Common Names: **WHITE BRITTLEBUSH (INCIENSO).**
Arizona and California deserts. *(Encelia farinosa)*. Yellow. November-**May.**
Sunflower family. Size: Perennial shrubs, 2 to 3 feet high.

These low, branching shrubs with gray-green leaves are common on rocky slopes and benches where they lighten the winter landscape with their bright flower heads and create a spectacular mass of bloom during early spring. Flower stems rise several inches above the brittle leaf-covered branches, thus hiding the plant under a blanket of blossoms at the height of the blooming period.

Plants are abundant on rocky slopes below 3,000 feet from southern Nevada to Lower California and eastward through Arizona.

Stems exude a gum prized as incense by the early-day Catholic priests. Indians chewed this gum, and also heated it to smear on their bodies for the relief of pain.

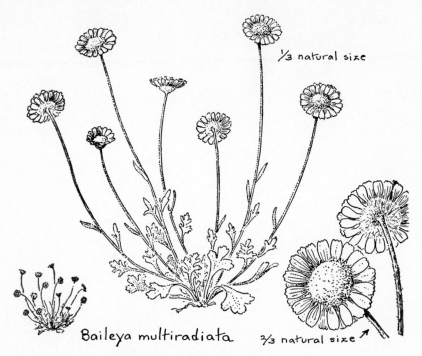

¹⁄₃ natural size

Baileya multiradiata ⅔ natural size

Common Names: DESERT BAILEYA (DESERT-MARIGOLD, WOOLLY-MARIGOLD, PAPERDAISY).

Arizona and Texas deserts. *(Baileya multiradiata).* Yellow. March-October.
California desert. *(Baileya pleniradiata).* Yellow. March-November.
Sunflower family. Size: 4 inches up to 2 feet high.

This low-growing, woolly, short lived perennial herb with showy, yellow flowers on long, solitary stems is one of the commonest bloomers gracing the desert roadsides and making patches of bright color along otherwise drab and dry, sandy washes. It is particularly noticeable because of its luxurious crop of flowers and long period of bloom.

At first glance, desert baileya may be confused with crownbeard, to which it is quite similar in color, size, and habit of growing in groups. However, the regular, circular shape of marigold blooms and the considerable difference in leaf shape make the two readily distinguishable.

In California, Desert baileya or "marigold" is cultivated for the flower trade.

Fatal poisoning of sheep on over-grazed ranges has been laid at the door of this plant, although horses crop the flower heads, apparently without harmful effect. Blossom petals (rays) become bleached and papery as the blossoms age, thus giving the plant in some localities the name "paperdaisy."

Desert baileya, of which there are but a few species, is common throughout desert areas of the Southwest from Utah and Nevada to Lower California, Sonora and Chihuahua.

YELLOW

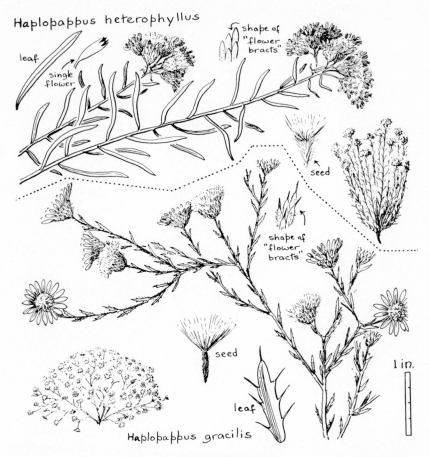

Haplopappus heterophyllus

leaf

single flower

shape of "flower bracts"

seed

shape of "flower bracts"

seed

leaf

1 in.

Haplopappus gracilis

Common Names: **GOLDENWEED, JIMMYWEED (RAYLESS-GOLDEN-ROD, GOLDENBUSH, LARCHLEAF GOLDENWEED).**

Arizona desert. *(Haplopappus lacrifolius).* Yellow. August-November.

California desert. *(Haplopappus gracilis).* Yellow. February-November.

Texas-New Mexico desert. *(Haplopappus heterophyllus).* Yellow. June-September.

Sunflower family. Size: Herbs or small shrubs 2 to 18 inches.

The genus *Haplopappus* (sometimes spelled *Aplopappus*) is represented in the Southwest by a great many species, both annuals and perennials, which range from elevations of 2,000 feet up to 9,000 feet. Desert forms prefer open, dry canyon slopes and mesas.

Haplopappus gracilis is not limited to lowland desert, but may be found on dry mesas and rocky slopes up to elevations of 6,000 feet.

Haplopappus heterophyllus often takes over heavily grazed range-land since it is generally unpalatable to livestock and replaces vegetation destroyed by overgrazing.

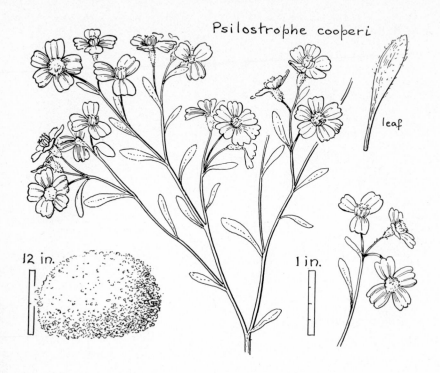

Psilostrophe cooperi

leaf

12 in.

1 in.

Common Name: **PAPERFLOWER**

Arizona and Texas deserts. *(Psilostrophe cooperi).* Bright yellow. Year around.

Texas-New Mexico deserts. *(Psilostrophe sparsiflora).* Bright yellow. May-September.

Sunflower family. Size: Rounded bush 12 to 18 inches high.

One man of the writer's acquaintance, confused by the great number of yellow flowers on the desert, refers to them all as "yellow composites." The paperflower is one of these.

It is noticeable because of the conspicuous, bright yellow flowers which sometimes cover the plants almost completely, often during periods of the year when bloom is quite scarce on the desert.

The flowers are persistent, petals (rays) become papery, fade to a pale yellow, and remain on the plants intact for weeks.

Although the paperflower does not form great masses of color, the blossom-covered clumps are conspicuous among the cactus, mesquite, and creosotebush of the desert.

It is common at elevations below 5,000 feet from southern Utah to Lower California, with similar species ranging eastward through southern New Mexico and northern Chihuahua.

Some species are reported to be poisonous to sheep.

YELLOW

Enceliopsis or Sunray

Joshua-tree in flower

Mexican Goldpoppy in full bloom

Spring bloom on the Sonoran Desert, Southern Arizona

Bladderpod, Desert Chicory, and Mexican goldpoppies

Fishhook Mammillaria

Brittlebush & Mallow

Brittlebush

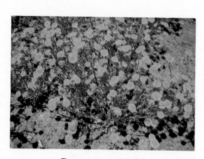

Desert-marigold

Walkingstick cholla

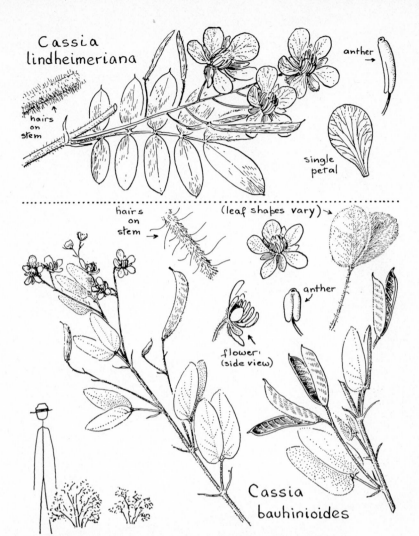

Cassia
lindheimeriana

anther

hairs
on
stem

single
petal

hairs
on
stem

(leaf shapes vary)

anther

flower
(side view)

Cassia
bauhinioides

Common Names: **SENNA, DESERT SENNA (RATTLEWEED).**
Arizona desert. *(Cassia bauhinioides).* Yellow. May-August.
California desert. *(Cassia armata).* Yellow. April-May.
Texas desert. *(Cassia lindheimeriana).* Golden. June-September.
Pea family. Size: Low, branching shrub up to 3 feet.

Members of this large genus are chiefly tropical, the majority
having golden to bronze flowers and brown, woody seed pods. They
are quite common along desert roadsides, and a few species are cultivated as ornamentals.

In some localities, following moist winters, desert senna bursts
into a riot of color in April and May adding a golden glory to the
spring floral display.

Representatives of the several desert species occur at elevations
between 2,000 and 5,000 feet from Texas westward to southern California and south into Mexico.

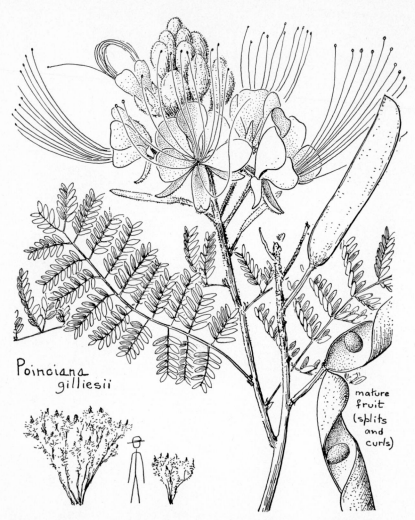

Poinciana
gilliesii

mature
fruit
(splits
and
curls)

Common Names: **PARADISE POINCIANA, (BIRD-OF-PARADISE).**
Arizona desert. *(Poinciana gilliesii).* Yellow-and-red. May-August.
Pea family. Size: Shrub, up to 10 feet tall.

Widely grown as a decorative shrub by the people of Mexico, this spectacular import from South America is quite commonly used as an ornamental in yards and around houses in desert areas of the Southwest. Under suitable conditions, it may escape and grow wild. The very showy blossoms with yellow petals and long, thread-like, red filaments are certain to attract attention.

In contrast to the striking showiness of the blossoms, the plant itself is straggling and unsymmetrical, and gives off an unpleasant odor.

YELLOW   58

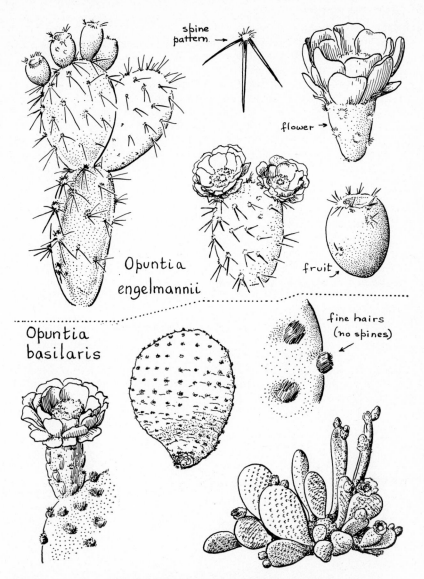

spine pattern →

flower →

Opuntia engelmannii

fruit →

Opuntia basilaris

fine hairs (no spines) →

Common Names: **PRICKLYPEAR, ENGELMANN PRICKLYPEAR, BEAVERTAIL PRICKLYPEAR (TUNA).**

Arizona desert. *(Opuntia engelmannii).* Yellow. April-June.

California desert. *(Opuntia basilaris).* Magenta. March-April.

Texas desert. *(Opuntia engelmannii).* Yellow. May-July.

Cactus family. Size: Clumps, sometimes 5 feet high and 10 feet in diameter.

The flattened pods, or stem joints, of the pricklypears growing, as they do, in huge clumps, make them the best known of the cactuses throughout the West. There are many species found throughout the

59   **YELLOW**

United States, but the plants reach their greatest size and most luxuriant growth in the desert areas of the Southwest. The large, red to purple and mahogany, juicy, pear-shaped fruits are known as *tunas,* and are eaten by many animals as well as by the native peoples. Flowers are large and spectacular.

Although a number of species of pricklypears are found in all of the desert areas, *Opuntia engelmannii* with its bright yellow flowers is the commonest form in both the Sonoran and Chihuahuan Deserts, while the beavertail pricklypear with its magenta flowers and lack of large spines is the common and spectacular form of the Mojave Desert.

Pricklypears are increasing in parts of the desert where conditions are favorable, especially where heavy grazing has given them an advantage over plants that are palatable to livestock.

Common Names: **BARRELCACTUS (COMPASS CACTUS, DEVILS-HEAD CACTUS, BISNAGA, BISNAGRE).**
Arizona desert. *(Ferocactus wislizenii).* Orange-yellow. July-September.
California desert. *(Ferocactus acanthodes).* Yellow. March-May.
Texas desert. *(Echinocactus horizonthalonius).* Rose-pink. May-June.
Cactus family. Size: 2 to 8 feet high.

Well known among the desert figures are the heavy-bodied barrelcactuses which are sometimes pointed out as sources of water for travelers suffering from thirst. Under extreme conditions, it is possible to hack off the tops of these tough, spine-protected plants and obtain, by squeezing the macerated tissues, enough juice to sustain life.

Growing faster on the shaded side, the taller-growing plants tend to lean toward the south, hence the name "compass" cactus. Flowers range in color from yellow to orange and rose-pink, depending on the

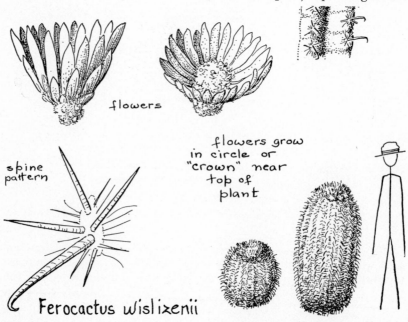

flowers

spine pattern

flowers grow in circle or "crown" near top of plant

Ferocactus wislizenii

species, and the pale yellow, egg-shaped fruits which ripen early in the winter, are a favorite food of deer and rodents. Flowers, and the resulting fruits, form a ring around the crown of the plant.

The flesh of the barrelcactus, cooked in sugar, forms a base of cactus candy.

Common Names: **AGAVE (CENTURYPLANT, MESCAL, LECHUGUILLA).**

Arizona desert. *(Agave palmeri).* Yellow-purple. July-August.
California desert. *(Agave desertii).* Yellow. May.
Texas-New Mexico desert. *(Agave lechuguilla).* Lavender-brown. April-May.
Amaryllis family. Size: Flower stalks 8 to 25 feet tall.

Many species of agave are found in various parts of the desert, hence it is difficult to settle on those which should be given particular recognition. Their blossoms, in general, are various shades of yellow. The larger species are called centuryplant or mescal (mess-KAHL), while the small ones are spoken of as lechuguillas (letch-you-GHEE-ahs). The lechuguilla, covering hundreds of square miles in Texas, New Mexico, and northern Mexico, is an indicator of the Chihuahuan Desert, holding the position in that desert which the

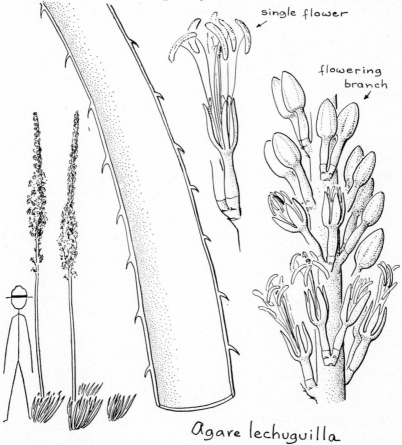

single flower

flowering branch

Agave lechuguilla

Saguaro does in the Sonoran Desert and the Joshua-tree in the Mojave Desert.

From its leaf fibers the Mexicans weave a coarse fabric. Its plumelike flower stalks, relished by deer and cattle, form one of the spectacular sights of the Chihuahuan Desert in springtime.

Agave plants require a number of years to store sufficient plant foods for the production of the huge flower stalk which grows with amazing rapidity to produce the many flowers and seeds, after which the plant dies. This long pre-blossom period of a dozen to 15 or more years is the basis for the name "centuryplant." If the young stalk is cut off, the sweet sap may be collected and fermented to form highly intoxicating beverages, some of which are distilled commercially. Among these are mescal, pulque (POOL-kay), and tequila (tay-KEEL-ah). Indians cut the young bud stalks, and roast them in rock-lined pits.

Common Names: **DESERT-MARIPOSA, MARIPOSA**
Arizona and California deserts. *(Calochortus kennedyi)*. Orange.
    March-May.
Lily family. Size: Perennial, about 2 feet tall.

Under favorable weather conditions, this short-stemmed mariposa presents a gorgeous display of spring color. Closely related to the white-flowered weakstem mariposa *(Calochortus flexuosus)* and to the sego lily (state flower of Utah), the desert-mariposa is found below 5,000 feet in Nevada, southern California, southern Arizona, and northern Sonora. When growing beneath taller shrubs, it forsakes its short-stemmed habit and forces its way up through the low branches, displaying its blossom above.

The mariposas, of which there are several species, are among the most beautiful wildflowers of the Southwest.

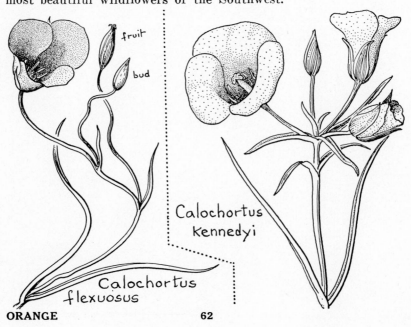

fruit

bud

Calochortus
kennedyi

Calochortus
flexuosus

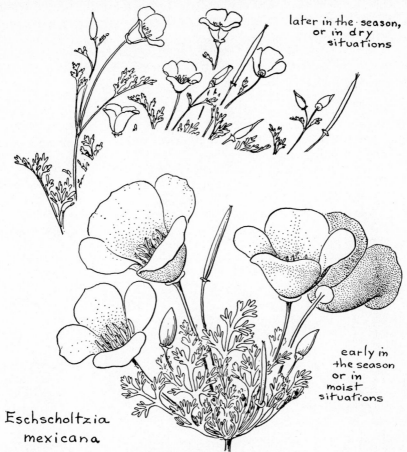

later in the season, or in dry situations

early in the season or in moist situations

Eschscholtzia mexicana

Common Names: **GOLDPOPPY, DESERT GOLDPOPPY, MEXICAN GOLDPOPPY (COPA DE ORO).**

Arizona desert. *(Eschscholtzia mexicana).* Orange. February-May.
California desert. *(Eschscholtzia glyptosperma).* Bright yellow. March-May.
Poppy family. Size: 3 inches to a foot high, with many flower stems.

Because of their abundance and dense growth, following winters of heavy precipitation, these annual poppies often cover portions of the desert with a "cloth of gold." They are closely related to the well-known California poppy, state flower of California, and a common border or bedding plant in home flower gardens. In the desert, goldpoppies are sometimes mixed with owlclover, lupines, and other spring flowers forming a multi-colored carpet that attracts visitors from great distances. (See cover.)

Common Names: **DEVILSCLAWS, UNICORNPLANT (ELEPHANT-TUSKS).**

Arizona desert. *(Proboscidea parviflora).* Orange-purple. April-October.
California desert. *(Proboscidea altheaefolia).* Coppery yellow. July-September.
Texas desert. *(Proboscidea arenaria).* Coppery yellow. July-September.
Unicornplant family. Size: Trailing, with stems 2 to 5 feet long.

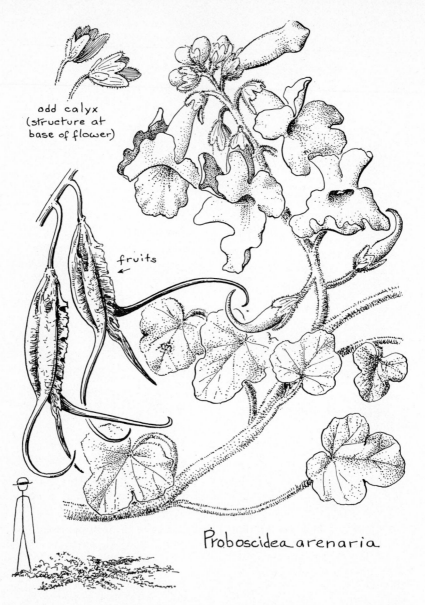

odd calyx
(structure at
base of flower)

fruits

Proboscidea arenaria

The showy, reddish-purple to coppery-yellow flowers, which are large enough to attract attention, are relatively few. More spectacular are the large, black, woody pods ending in two curved, prong-like appendages that hook about the fetlocks of burros or the fleece of sheep, thereby carrying the pod away from the mother plant and scattering the seeds. Young pods are sometimes eaten by desert Indians as a vegetable, and the mature fruits are gathered by the Pima and Papago Indians, who strip off the black outer covering and use it in weaving designs into basketry.

COPPERY   64

Common Names: **BURROBRUSH (CHEESEWEED).**

Arizona and California deserts. *(Hymenoclea salsola).* Silvery red.
   March-April.

Texas-New Mexico deserts. *(Hymenoclea monogyra).* Silvery red. September.

Sunflower family. Size: Much-branched shrub, 2 to 3 feet tall.

    Burrobrush is another of the common desert shrubs whose fruits
are much more conspicuous than the blossoms. The shrub itself is
bright green in color, and somewhat resembles the common Russian-
thistle. It is widespread, and abundant in sandy washes, where it
tends to form thickets.

    In some localities it is called "cheeseweed" because of the cheesy
odor of the crushed foliage.

    It occurs throughout the Southwest at elevations below 4,000
feet, from western Texas to southern California and northern Mexico.

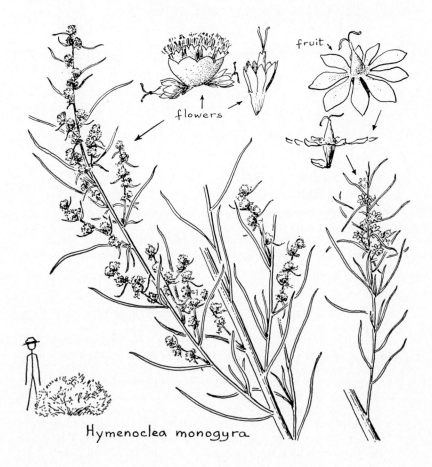

fruit

flowers

Hymenoclea monogyra

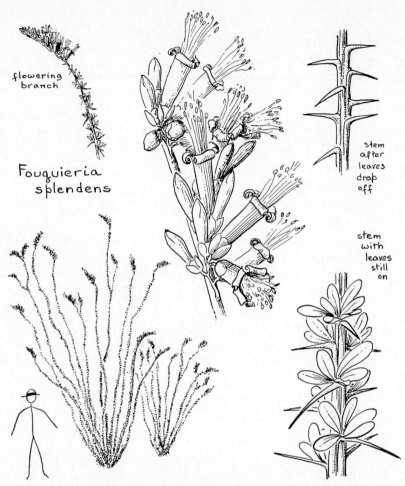

flowering branch

Fouquieria splendens

stem after leaves drop off

stem with leaves still on

Common Names: OCOTILLO (SLIMWOOD, COACHWHIP, CANDLE-WOOD, FLAMINGSWORD).

Arizona, California and Texas deserts. *(Fouquieria splendens).* Bright red. April-May.

Ocotillo family. Size: Up to 15 feet tall.

One of the few flower families restricted to the desert, the unique ocotillo (oh-ko-TEE-oh) with its long, unbranching stems is found on rocky hillsides below 5,000 feet from western Texas to southern California and south into Mexico. It is one of the commonest, queer-est, and most spectacular of desert plants, especially when the tips of its long, slender stems seem afire with dense clusters of bright red blossoms. Following rains, leaves clothe the thorny stems with green, but after the soil becomes dry, the leaves turn brown and fall. The heavily thorned stems are covered with green bark which takes over the functions of leaves during periods of drought. The plant thus becomes semi-dormant during hot dry periods and, in sections of the desert visited by showers, may go through this cycle several times during a year.

Because of its sharp thorns, the ocotillo is thought by many strangers to the desert to be one of the cactuses. It is much more closely related to both the primrose family and the olive family than to the cactus family.

Stems of the ocotillo are used by natives in building huts. They are sometimes cut and, when planted close together in rows, take root and form living fences and corrals.

Common Names: **TRAILING ALLIONIA (WINDMILLS, PINK THREE-FLOWER, TRAILING FOUR-O'CLOCK).**
Arizona and Texas deserts. *(Allionia incarnata)*. Purple pink. April-October.
California desert. *(Allionia albida)*. Rose pink. July-October.
Four-o'clock family. Size: Spreading annual with branches 30 inches.

Slender, trailing stems up to 30 inches in length with clusters of three rose-purple to pink blossoms serve to identify the allionia which is a conspicuous plant of the open plains and mesas. The plants prefer dry, sandy benches where they are quite conspicuous with their prostrate, somewhat sticky stems weighted with clinging

"flower" really three separate flowers

Allionia incarnata

×7

fruit

grains of sand. Blossoms are usually showy and colorful, rarely pale rose to white.

Fruits of *Allionia incarnata* are conspicuously toothed.

———

Common Names: **GLOBEMALLOW (APRICOT-MALLOW, SORE-EYE POPPY, DESERTMALLOW).**
Arizona and California deserts. *(Sphaeralcea ambigua).* Peach-pink. February-May.
Texas-New Mexico deserts. *(Sphaeralcea angustifolia).* Pink. May-October.
Mallow family. Size: 1 to 5 feet tall, often clustered.

Common throughout all of the Southwest, the mallows range in size from small herbs 5 or 6 inches high to coarse, straggling, woody-stemmed plants with stems 4 or 5 feet long. Their flowers range in color from white and pale yellow to lavender, apricot, and red. Some species, including *ambigua,* grow in large clumps with as many as 100 stems from a single root. The smaller species often cover the desert floor in early spring with a dense growth of flowers giving an apricot tinge to the landscape. Several species flower in spring and again after the summer rains.

A local belief that hairs of the plant are irritating to the eyes has given the name "sore-eye poppies," an appellation carried out in the Mexican name *mal-de-ojos.* In Lower California, mallows are called *plantas muy malas,* meaning very bad plants. In contrast, the Pima Indian name is translated to mean "a cure for sore eyes."

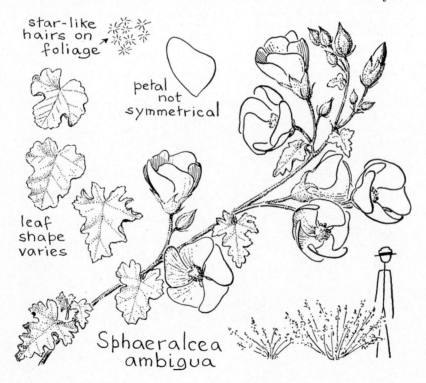

star-like hairs on foliage

petal not symmetrical

leaf shape varies

Sphaeralcea ambigua

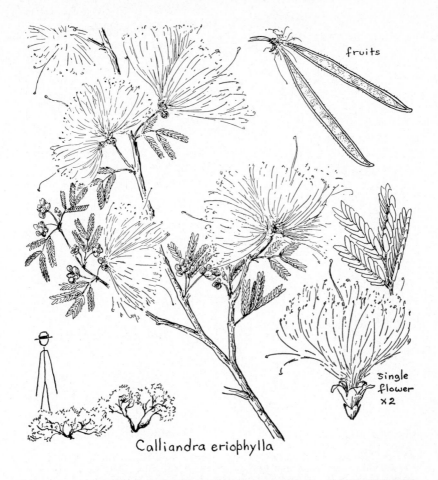

Calliandra eriophylla

fruits

Single flower x2

Common Names: **FALSE-MESQUITE CALLIANDRA (FAIRYDUSTER, CALLIANDRA, MESQUITILLA).**

Arizona, California and Texas deserts. *(Calliandra eriophylla).* Pink. February-May.

Pea family. Size: From a few inches to 3½ feet tall.

This straggling, perennial shrub with fine, mimosa-type leaves is common over much of the desert, lining banks of arroyos or dotting open hillsides. It is particularly conspicuous when in flower because of the spectacular tassel-like blossoms which are white and scarlet, or generally pink in appearance. The small leaves are nutritious and are highly palatable to deer and to livestock. The petite fairyduster adds much to the color and springtime atmosphere of the desert. It is particularly noticeable along the base of the Tanque Verde hills in Saguaro National Monument.

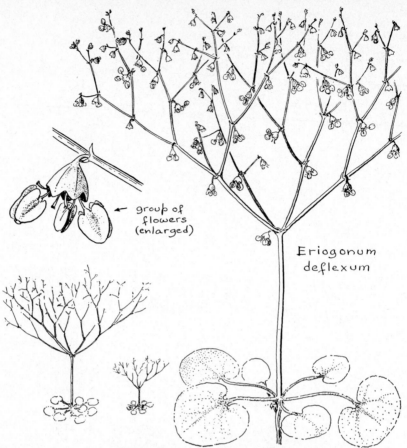

group of
flowers
(enlarged)

Eriogonum
deflexum

Common Names: **ERIOGONUM (SKELETONWEED, DESERT BUCKWHEAT).**

Arizona desert. *(Eriogonum densum).* Pink. May-October.
California desert. *(Eriogonum deflexum).* Pink-white. All year.
Texas desert. *(Eriogonum polycladon).* Pink. June-November.
Buckwheat family. Size: 6 inches to 30 inches high.

Eriogonum is a very large genus, many species of which are common, and contains both annuals and biennials. Although the flowers are small, they are usually numerous and conspicuous. *Eriogonum densum* is often very abundant in semi-desert areas, particularly along roadsides, where it is especially noticeable because it monopolizes the pavement edges for miles. It is extremely resistant to drought and flourishes when many other herbaceous plants have dried out completely. Although it bears flowers at almost any time throughout the year, during the autumn months the branches are loaded with myriads of pendant, pearly flowers the size of rice kernels. In winter, the stalks turn maroon in color and are quite conspicuous.

Eriogonum polycladon is often so common along roadsides and desert washes as to color the landscape with its grayish stems and pink flowers.

Eriogonum inflatum always attracts attention because of its swollen stems which resemble tall, slender bottles.

PINK

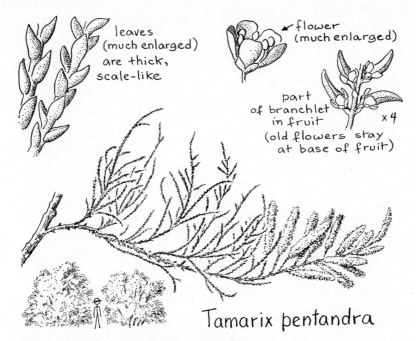

leaves (much enlarged) are thick, scale-like

flower (much enlarged)

part of branchlet in fruit (old flowers stay at base of fruit)

x 4

Tamarix pentandra

Common Names: **TAMARISK (SALTCEDAR).**
Arizona, California, and Texas deserts. (*Tamarix pentandra*). Pink to white.
  March-August.
Tamarix family. Size: Shrubs to trees up to 15-20 feet high.

Purists could object to inclusion of the tamarisk in this booklet because it is not native. However, due to a number of importations (eight species being introduced by the Department of Agriculture between 1899 and 1915) and to its ability to spread rapidly under suitable conditions, it is now widespread throughout the Southwest.

Tamarisk grows as a graceful shrub or small tree with drooping branches covered with small, scale-like leaves and is abundant in moist locations below 5,000 feet. It prefers a hot climate, low humidity, and saline soils. In river bottoms it often forms dense thickets which require immense quantities of water, hence rob the few desert streams of a high percentage of their moisture.

Honeybees obtain nectar from the blossoms, which are particularly noticeable in the spring and early summer, completely covering the branches which appear as light pink, drooping plumes. The thickets are valuable as windbreaks and in erosion control. Once established, they are very difficult to control, and because of the deep shade cast by their dense growth and the heavy feeding of the shallow roots, they prevent cropping.

The name tamarisk is often confused with the name of the larch or tamarack tree. There is little similarity except in the name.

The larger *Tamarix aphylla* (athel tamarisk) is similar in appearance but much larger and suitable for cultivation as a shade and decorative tree. It is subject to winterkill, but does not have the bad habit of spreading, characteristic of *Tamarix pentandra*.

PINK

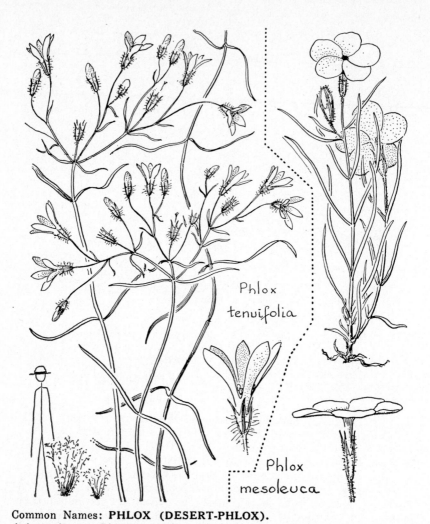

Phlox
tenuifolia

Phlox
mesoleuca

Common Names: **PHLOX (DESERT-PHLOX).**
Arizona desert. *(Phlox tenuifolia).* White-lavender. Spring.
California desert. *(Phlox stansburyi).* Pinkish-red. May-July.
Texas desert. *(Phlox mesoleuca).* Pink-white. June-August.
Phlox family. Size: Low-growing perennials, in clumps; or shrubby plants
   in tufts up to 3 feet tall.

   Representatives of the phlox genus are found from the hot desert
lowlands to the mountain tops well above the timberline. Certain
species are limited in their range to the desert areas of the Southwest, and it is in these that we are interested here. The plants sometimes present a mass of heavy bloom twice yearly; heaviest in the
spring, and again following the summer rains. Several of the native
species have been brought under cultivation, particularly *Phlox
tenuifolia,* in desert gardens, as it grows naturally in a bushy habitat
similar to that formed by the shrubs planted around a house. Other
forms grow as low, creeping mats forming fragrant, colorful floral
carpets.

LAVENDER   72

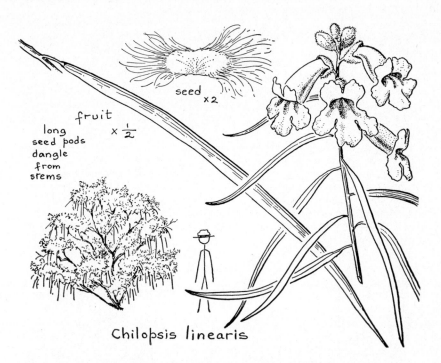

fruit
x ½

long
seed pods
dangle
from
stems

seed x2

Chilopsis linearis

Common Names: DESERTWILLOW (DESERT-CATALPA).

Arizona, California, and Texas deserts. *(Chilopsis linearis).* Pink-lavender. April-August.

Bignonia family. Size: Shrubby tree, 6 to 15 feet high.

Although a close relative of the catalpa, the willow-like foliage of this small tree has given it the name desertwillow. A small and inconspicuous part of the desert vegetation when not in flower, unnoticed among the heavier growth of trees and shrubs that crowd the banks of desert washes, the tree's beautiful orchid-like flowers of white to lavender mottled with dots and splotches of brown and purple bring exclamations of delight from persons viewing them for the first time. Because of the beauty of the tree when in bloom, it is sometimes cultivated as an ornamental.

Leaves are rarely browsed by livestock, and the durable, black-barked wood is used for fenceposts. In Mexico, a tea made by steeping the dried flowers is considered to be of medicinal value. By early autumn, the violet-scented flowers which appear after summer rains are replaced by the long, slender seed pods which remain dangling from the branches and serve to identify the tree long after the flowers are gone.

Although desertwillows are never found in pure stands, growing singly and rather infrequently among other trees and shrubs lining desert washes, the species is quite common below 4,000 feet across the entire desert from western Texas to southern Nevada, southern California and southward into Mexico.

LAVENDER

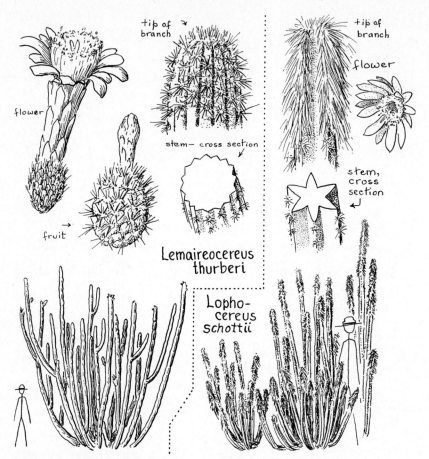

Lemaireocereus thurberi

Lopho-
cereus
schottii

Common Names: **ORGANPIPE CACTUS, SENITA (PITAHAYA DULCE).**

Arizona desert. *(Lemaireocereus thurberi).* Pink lavender. May-June.
Arizona desert. *(Lophocereus schottii).* Pink. April-August.
Cactus family. Size: In clumps, stems up to 15 feet.

Two somewhat similar, columnar cactuses occur in the United States only in Organ Pipe Cactus National Monument and in its immediate vicinity. Both are fairly common in northwestern Mexico.

These two spectacular desert giants with their clumps of erect branches are sufficiently similar to be readily confused at first glance. However, the stems of the Organpipe *(Lemaireocereus thurberi)* are longer and contain more but much smaller ridges than do the stems of the senita or "whisker cactus." The name "senita" (meaning old age) refers to the long, gray, hair-like spines covering the upper ends of the senita stems.

Both species are night-blooming, the flowers, which appear along the sides and at the tips of the stems, closing soon after sunrise the following morning. Fruits of the organpipe are harvested by the Papago Indians.

LAVENDER   74

Although these two species of cactus are restricted to a very limited area, they are sufficiently spectacular and interesting to be considered worthy of inclusion in this booklet. It was to protect these species, threatened with extinction in the United States, and other rare and interesting forms of desert plants and animals, that Organ Pipe Cactus National Monument was established.

Common Names: MAMMILLARIA, FISHHOOK MAMMILLARIA, (PINCUSHION CACTUS, CORKSEED CACTUS, NIPPLE CACTUS, BUTTON CACTUS).

Arizona desert. *(Mammillaria microcarpa).* Lavender. June-July.

California desert. *(Mammillaria tetrancistra).* Lavender. June-July.

Texas-New Mexico deserts. *(Mammillaria micromeris).* Lavender. **Early summer.**

Cactus family. Size: Cucumber-shaped and 3 to 10 inches high.

Unlike blossoms of many of the cacti, flowers of the little *Mammillarias* often last for several days. Blossoms are pink or lavender, occasionally yellow, while the fruits are finger- or club-shaped and red. Being small and forming low clumps, or with single pincushion-like stems, they often escape attention except when glorified with bright, comparatively large flowers, which frequently form a crown around the top of the plant. The long spines are curved at the tips giving the plant the appearance of being covered with unbarbed fishhooks.

The mammillaria cactuses, of which there are a number of species throughout the Southwest, occur in dry, sandy hills from southern Utah to western Texas and in southern California and northern Mexico. The red fruits are bare, without scales, spines, or hairs.

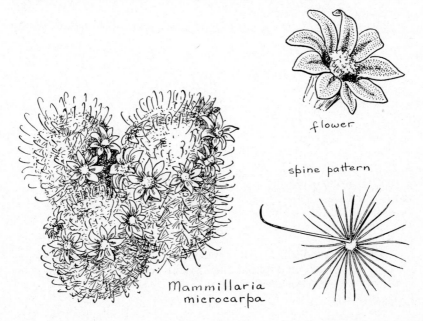

flower

spine pattern

Mammillaria
microcarpa

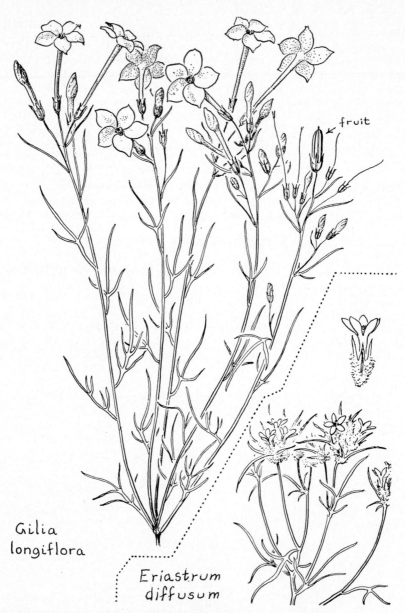

Gilia
longiflora

fruit

Eriastrum
diffusum

Common Names: **GILIA (STARFLOWER).**
Arizona desert. *(Eriastrum diffusum).* Lavender. April-May.
California desert. *(Gilia latifolia).* Pink-lavender. March-April.
Texas desert. *(Gilia longiflora).* Blue-lavender. April-October.
Phlox family. Size: 6 to 24 inches high.

    Although the gilias are not generally well known, they are common, quite widely distributed throughout the Southwest, and their beauty deserves wider recognition. There are a great many species

LAVENDER   **76**

(of which early flowering *Gilia inconspicua* is perhaps the common-est) at higher elevations as well as throughout the desert. Many of these are worthy of cultivation as ornamentals. Desert species, in general, are pale blue, white, or lavender while those of the higher elevations are pink, coral, or yellow to scarlet; although this is by no means a hard-and-fast rule.

Following winters of above-normal precipitation, desert species sometimes produce such heavy stands that the flowers cover large areas with a delicate pale blue or lavender carpet. Some species are attractive to hummingbirds.

Common Names: **PHACELIA (SCORPIONWEED, WILD-HELIOTROPE).**

Arizona desert. *(Phacelia crenulata).* Violet-purple. February-June.
California desert. *(Phacelia distans).* Blue-violet. March-May.
Texas desert. *(Phacelia coerulea).* Violet-purple. March-April.
Waterleaf family. Size: 4 to 16 inches tall.

Although strongly scented, it is not accurate to refer to these annuals as fragrant, for they are sometimes unpleasant in odor, and occasionally actually foul-smelling. Some are described as having an onion-like odor.

Phacelia crenulata with its rich, violet-purple flowers is conspic-uous across southern New Mexico, Arizona and California to Lower California. This species is often called wild-heliotrope.

The name scorpionweed comes from the curling habit of the blossoming flower heads which somewhat resemble the flexed tail of a scorpion in striking position.

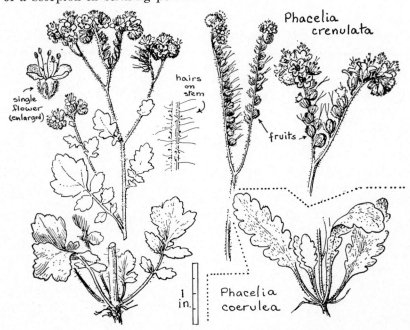

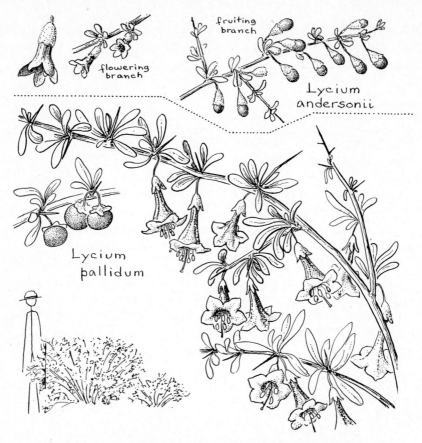

flowering branch

fruiting branch

Lycium andersonii

Lycium pallidum

Common Names: **WOLFBERRY (SQUAW-THORN, RABBIT-THORN, DESERT-THORN, SQUAWBERRY, TOMATILLO).**

Arizona desert. *(Lycium pallidum)*. Green-lavender. April-June.

California desert. *(Lycium andersonii)*. Lavender. February-April.

Texas desert. *(Lycium berlanderi)*. Lavender-cream. March-September.

Potato family. Size: Thorny shrubs, stiff and brushy, up to 6 feet.

Noticeable in winter because of their off-season greenery and early flowers which cover the bushes and attract many insects, and attractive in late spring and summer due to the numerous tomato-colored berries hanging from their stiff, thorny stems, the wolfberry is widely distributed throughout the desert.

These plants have contributed much to the subsistence of the Indians, their insipid, slightly bitter, juicy berries being eaten raw or prepared as a sauce. These berries are eagerly sought by birds, which also use the stiff shrubs for cover and for protective roosts at night.

Early spring is the normal blooming season, but some flower again following summer or early fall rains.

VIOLET

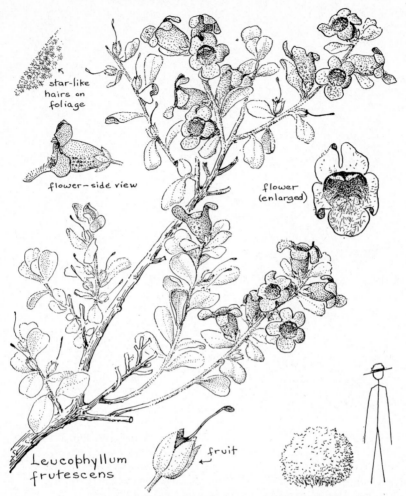

star-like hairs on foliage

flower – side view

flower (enlarged)

Leucophyllum frutescens

fruit

Common Names: **SILVERLEAF (CENIZA, SENISA, ASHPLANT, WILD-LILAC).**

Texas desert. *(Leucophyllum frutescens)*. Lilac-violet. August-October.
        (Leucophyllum texanum). Violet-purple. August-October.
Figwort family. Size: Bushy shrub, 3 to 4 feet high.

In southern Texas, thick patches of this shrub are sometimes found, although they commonly occur singly or a few together, usually on limestone soils. Since the leaves are a light gray-green, plants appear to be ashy in color, giving rise to the name "ceniza," meaning "ashy." Spectacular in Big Bend National Park.

So sensitive is this plant to moisture, that it may burst into blossom within a few hours after soaking rain, this phenomenon giving rise to the local belief that the plant actually blossoms before the rain, thereby forecasting precipitation; hence it is sometimes called "barometerbush." During recent years silverleaf has become one of the popular native shrubs used in landscaping.

Under normal conditions, plants blossom in September.

      VIOLET

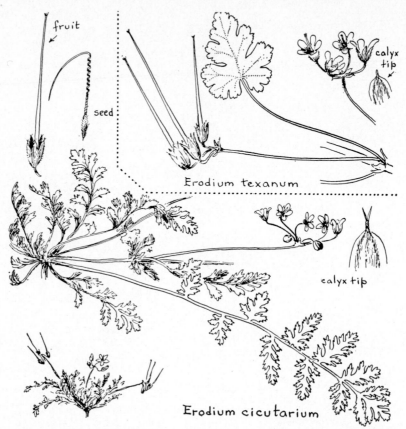

fruit

seed

calyx tip

Erodium texanum

calyx tip

Erodium cicutarium

Common Names: **HERONBILL, ALFILERIA (FILEREE OR FILAREE).**

Arizona and Texas deserts. *(Erodium texanum)*. Pink-violet. February-March.

California desert. *(Erodium cicutarium)*. Rose-violet. February-March.

Geranium family. Size: 3 to 12 inches high.

Believed to have been introduced from the Mediterranean countries at an early date by the Spaniards, alfileria is now widespread and extensively naturalized through the Southwest. In the desert, it is one of the common winter annuals and furnishes excellent spring forage especially following moist winters. The plants remain green for only a few weeks, but are good forage even after the stems have dried.

Although the blossoms are not large nor sufficiently numerous to make a colorful display, they are attractive and welcome, as they are among the first spring flowers to put in an appearance. "Tails" of the fruits are long and slender, somewhat resembling a heron's bill, and upon maturity twist into a tight spiral when dry. Upon becoming moist, they uncoil, driving the sharp-tipped seeds into the soil. Seeds are gathered and stored by ants which discard the husks and coiled "tails" outside their nests, thus building up a circular band of chaff around the anthill.

VIOLET 80

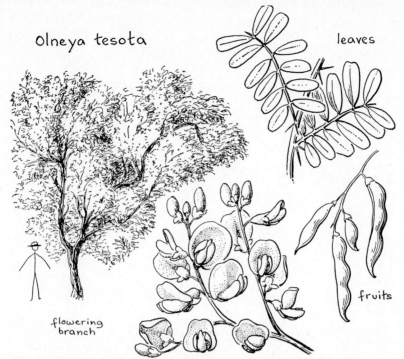

Olneya tesota

leaves

fruits

flowering
branch

Common Names: **TESOTA (IRONWOOD, DESERT-IRONWOOD, PALO-DE-HIERRO).**
Arizona and California deserts. *(Olneya tesota).* Violet-purple. May-June. Pea family. Size: Wide-crowned tree up to 35 feet.

Tesota is one of the desert's most beautiful trees, being particularly colorful when the new, dark-green leaves and violet, wisteria-like flowers give it a lavender glow in late May or early June. Since the tree survives only in warm locations, it has for years served as a guide to citrus growers in selecting sites for orange, lemon, or grapefruit plantings.

Foliage of the "ironwood" is dense and evergreen, and the wood is very heavy and so hard that it cannot be worked with ordinary tools. When thoroughly dry, it makes high-quality firewood, and as a result has been cut and removed over much of the desert, hence mature trees are becoming relatively scarce. Indians used the wood for arrow points and as tool handles.

The trees grow along desert washes, often in company with mesquite and paloverde. Blossoms are much more numerous in some years than in others. Although the trees, when in bloom, make a spectacular showing, they are very difficult to capture on color film, and photographs that do them justice are rare. Seeds, which mature late in the summer, are roasted and eaten by desert Indians who prize them for their peanut-like flavor. They are eaten also by various desert animals.

In Organ Pipe Cactus National Monument and in some other parts of the desert, "ironwood" trees have become heavily infested with mistletoe which stunts or kills the branches and produces grotesque, tumor-like swellings.

VIOLET

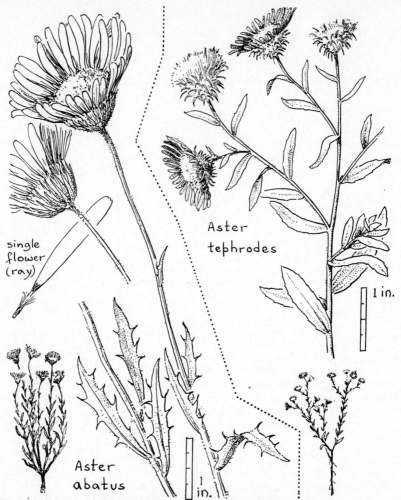

single
flower
(ray)

Aster
tephrodes

1 in.

Aster
abatus

1
in.

**Common Names: ASTER, MOJAVE-ASTER, TANSYLEAF ASTER
(DESERT ASTER).**

Arizona desert. *(Aster tephrodes).* Amethyst blue. April-October.
California desert. *(Aster abatus).* Violet to lavender. March-May.
Texas desert. *(Aster tanacetifolius).* Bright violet. June-October.
Sunflower family. Size: Few inches to 2½ feet tall.

Aster tanacetifolius
leaf
shapes

Since the aster is one of the most widespread and
best-known of the flowers, it is usually easily recognized. There are many species, principally perennials,
ranging from low-growing, single-stemmed plants,
sprawling, many-stemmed plants with large flowers,
to tall bushes. Desert species are found on dry, rocky
hillsides and along roadsides and on waste ground.

The aster is by no means restricted to the desert.
Over much of the United States they are considered
as fall bloomers, but many species blossom in the
spring while others are at their floral best in mid-
summer.

VIOLET

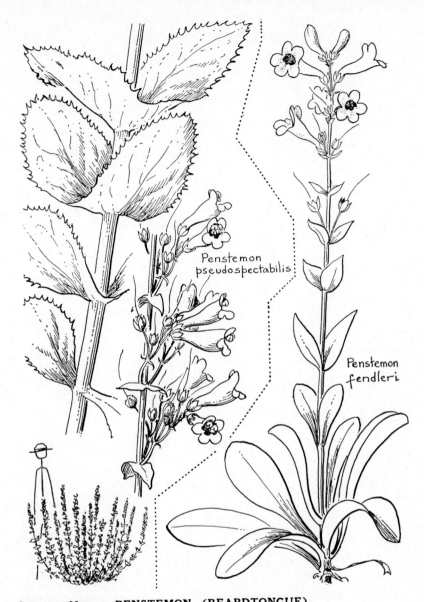

Penstemon
pseudospectabilis

Penstemon
fendleri

Common Names: **PENSTEMON (BEARDTONGUE).**

Arizona desert. *(Penstemon pseudospectabilis)*. Rose-purple. April-July.

California desert. *(Penstemon thurberi)*. Blue-purple. April-June.

Texas desert. *(Penstemon fendleri)*. Blue-purple. April-June.

Figwort family. Size: Perennial herbs from a few inches high to 3 feet or more tall.

   Widespread through the Southwest at nearly all elevations, the penstemons are conspicuous herbs or small shrubs with showy flowers that attract attention and admiration when they are in bloom in the spring and early summer on the desert.

PURPLE

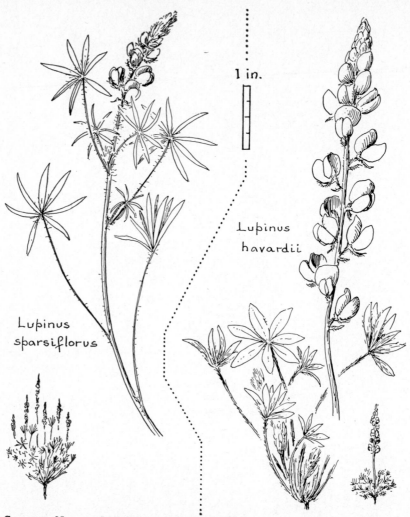

Lupinus
havardii

Lupinus
sparsiflorus

Common Names: LUPINE (BLUEBONNET).
Arizona desert. *(Lupinus sparsiflorus).* Violet-purple. January-May.
California desert. *(Lupinus odoratus).* Royal purple. April-May.
Texas desert. *(Lupinus havardii).* Blue-purple. March-April.
Pea family. Size: Bushy, and up to 2 or 2½ feet tall.

Lupines are among the old dependables of spring display flowers of the desert, usually mingling with other blossoming herbs to create the bright color pattern for which the desert is famous in early spring, but occasionally growing in pure stands. Ranging in color from pale pink to deep purple, the lupines are usually considered as blue flowers.

The name "lupine" comes from the Latin word meaning wolf and was applied to these plants because they were believed to rob the soil of its fertility. Actually, they prefer the poorer, sandy soils and, by fixing in the soil nitrogen that they, in common with other plants of the pea family, are able to obtain from the air, they actually improve the land on which they grow.

PURPLE   84

Perhaps the best known display of lupines takes place each spring in Texas. Here the "bluebonnet" *(Lupinus texensis* and *Lupinus subcarnosus)* has been named the state flower of Texas, and the annual spring display attracts thousands of people to the areas of heavy bloom. The majority of lupines have handsome flowers, some species are fragrant, and several species are cultivated as ornamentals. The seeds of a few species contain alkaloids which are poisonous to livestock, especially sheep.

Common Names: **BROOMRAPE (BURROWEED STRANGLER, CANCER-ROOT).**
Arizona, California, and Texas deserts. *(Orobanche ludoviciana).* Brownish-purple. March-July.
Broomrape family. Size: 4 to 15 inches tall.

This root parasite, although not common, is sufficiently strange and striking in appearance to arrest attention. Its purple to yellowish-brown, leafless flower stalks somewhat resembling coarse shoots of asparagus rise above the desert soil, usually in open, sandy locations.

Broomrape, of which there are several species, is found throughout the Southwest from southern Utah and Nevada to Texas, California, and Mexico.

The plant is parasitic on the roots of a number of different plants, but the desert species usually parasitize burrobush, bursage, and other composites. Flowers are small, purple with brown and white markings, and monopolize the plant stalk in the absence of foliage.

Underground parts of the plant were eaten by Southwestern Indians. The name "cancer-root" refers to the reported efficacy of treatment in applying the stems of the plant to ulcers.

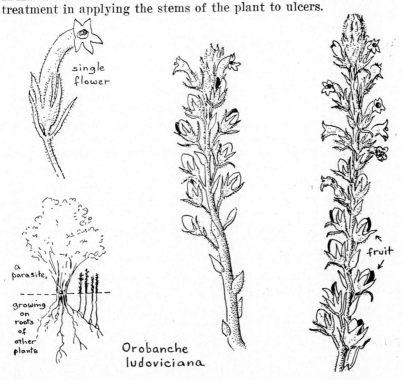

single flower

a parasite, growing on roots of other plants

Orobanche ludoviciana

fruit

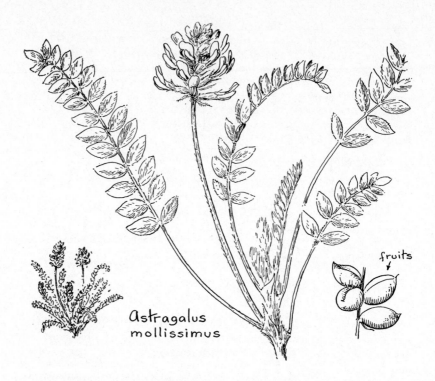

Astragalus
mollissimus

fruits

Common Names: **LOCO, WOOLLY LOCO, (MILKVETCH, LOCOWEED, RATTLEWEED).**

Arizona and California deserts. *(Astragalus nuttallianus).* White-purple. February-May.

Texas deserts. *(Astragalus mollissimus).* Purple. April-May.

Pea family. Size: 4 to 12 inches high.

A very large genus of plants, with 78 species recorded in Arizona alone, *Astragalus* ranges from the driest, hottest parts of the desert to high mountain peaks and the far north. *Astragalus nuttallianus* is the commonest of the desert species and is found on dry plains, mesas, and slopes below 4,000 feet from Arkansas and Texas westward to California and south into Mexico.

Some of the species, of which woolly loco *(mollissimus)* is one, contain selenium, a poisonous constituent causing the well-known and often fatal loco disease of livestock, particularly horses. (Loco is a Spanish word meaning "crazy.") Other species which prefer soils rich in selenium take up enough of that toxic mineral to make them poisonous to livestock, especially sheep.

Nearly all of the species are colorful and spectacular when in blossom, and some of them have a rank, disagreeable odor.

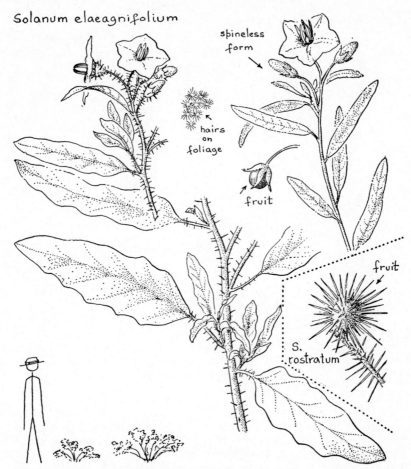

Solanum elaeagnifolium

spineless
form

hairs
on
foliage

fruit

fruit

S.
rostratum

Common Names: **NIGHTSHADE (GROUNDCHERRY, WILD POTATO, TROMPILLO, HORSENETTLE).**

Arizona and California deserts. *(Solanum xantii).* Purple. April-August. Texas desert. *(Solanum elaeagnifolium).* Purple-violet. May-September. Potato family. Size: Up to 3 feet.

Quite showy when in flower, these common roadside plants attract considerable attention during the late spring and summer. Some species become troublesome in cultivated fields and are difficult to eradicate. An alkaloid, solanin, reported as present in the leaves and unripe fruits of several species, renders them poisonous. Pima Indians add the crushed berries of *Solanum elaeagnifolium* to milk in making cheese.

The yellow-flowered *Solanum rostratum* is heavily covered with spines, including both stems and fruit, giving it the name of "buffalobur." This species is said to be the original host of the now widespread pest, the Colorado potato beetle.

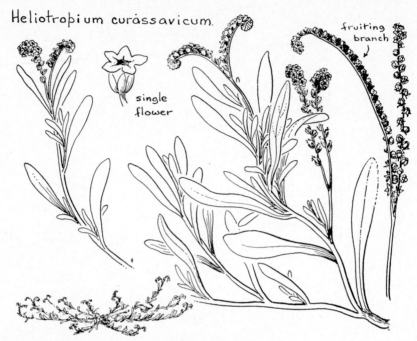

Heliotropium curassavicum.

single flower

fruiting branch

Common Names: **SALT HELIOTROPE (WILD-HELIOTROPE, QUAIL-PLANT, CHINESE-PUSLEY).**

Arizona, California and Texas deserts. *(Heliotropium curassavicum).* Purple. March-April.

Borage family. Size: Spreading weak stems up to 18 inches.

Widely distributed on salty and alkaline soils throughout the warmer parts of the Western Hemisphere, there are several species and varieties of wild heliotrope. The flowers, which are almost white, shading to a pale purple in the corolla throat, open as the spike uncoils, perfuming the desert air with their fragrance. The name "pusley" which is applied to this plant in some localities is possibly a corruption of "purslane."

Pima Indians are reported to powder the dried roots of these plants, applying the dust to wounds or sores. The name "wild heliotrope" is also applied to another desert flower, *Phacelia crenulata* (which see), causing no little confusion.

———

Common Names: **ARROWWEED PLUCHEA (MARSH-FLEABANE).**

Arizona, California, and Texas deserts. *(Pluchea sericea).* Roseate purple. Spring.

Sunflower family. Size: Perennial, 3 to 10 feet tall.

Seldom found above 3,000 feet elevation, the rank-smelling arrowweed pluchea forms dense, willow-like thickets in streambeds and in moist, saline soils. It is common in moist locations from Texas to southern Utah, and south into California and Mexico; usually in pure, dense stands.

PURPLE

88

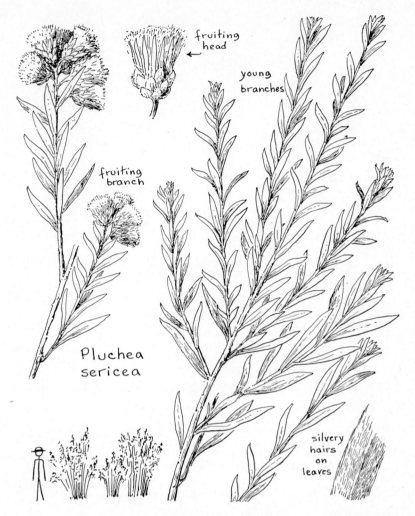

fruiting
head

young
branches

fruiting
branch

Pluchea
sericea

silvery
hairs
on
leaves

The green foliage gives off an agreeable odor, but when the plant dries this becomes rank and unpleasant, clinging to the plant long after it has been cut. This odor is often a characteristic of native dwellings where arrowweed has been used as a ceiling mat above the rafters.

Arrowweed is browsed by deer, and sometimes by horses and cattle. The straight stems were used by Indians in making arrow-shafts, and are still important as a construction material in the walls and roofs of mud huts. The stems are used, also, by desert Indians in basketmaking, and in fabricating storage bins and animal cages. From the foliage of the stem tips, Pima Indians brewed a tea which they used as an eye wash.

The flowers are reported to furnish considerable nectar gathered by honeybees. The blossoms are inconspicuous and develop into tawny-tufted seed heads.

PURPLE

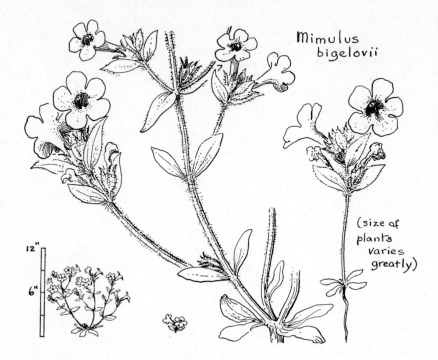

Mimulus bigelovii

(size of plants varies greatly)

12"

6"

Camon Name: **MONKEY-FLOWER**

Arizona and California deserts. *(Mimulus bigelovii).* Red-purple. February-April.

Texas desert. *(Mimulus glabratus).* Yellow. June.

Figwort family. Size: Branching, creeping annual up to 8 inches.

Disproportionately large flowers for the size of the low-growing, small-leafed plant make it particularly conspicuous in the open, sandy locations where it blossoms in the springtime.

Although the monkey-flower is usually thought of as moisture-loving, there are a number of desert species. The flowers are quite easy to recognize, as they closely resemble the monkey-flowers which grow in the moist places surrounding seeps and springs, and they also are somewhat similar in appearance to their close relatives, the snapdragons and penstemons.

The desert species are well worthy of consideration for cultivation as garden ornamentals.

leaves in basal rosette
and at ends of stems

stems
prostrate

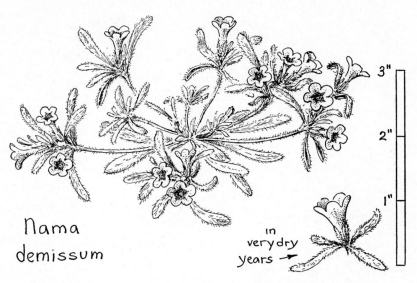

Nama
demissum

in
very dry
years →

3"

2"

1"

Common Names: **NAMA (PURPLEMAT, PURPLE ROLL-LEAF).**

Arizona, California and Texas deserts. *(Nama demissum).* Red-purple.
March-May.

Waterleaf family. Size: Tiny plant, an inch or so high.

Although the plants are very small, they grow close together and
the blossoms are often quite large in comparison. The reddish-
purple color of the flower stands out in sharp contrast to the green
of spring vegetation so that a widespread growth of the plants
forms patches or mats of colorful desert carpeting.

Masses of the plants are usually found on open flats, often
among creosotebush, and on either clay or sandy soils. In dry years,
growth is restricted and a tiny plant may bear but a single flower,
the blossom sometimes almost as large as the rest of the plant.

PURPLE

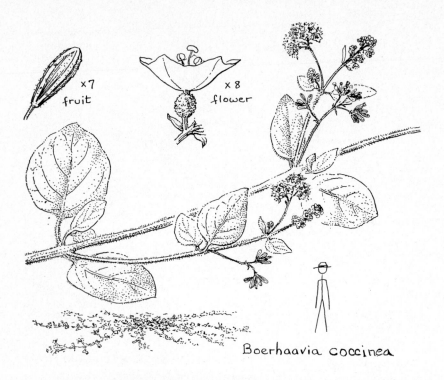

x7
fruit

x8
flower

Boerhaavia coccinea

Common Names: **SPIDERLING, SCARLET SPIDERLING (WEST INDIAN BOERHAAVIA).**

Arizona, California and Texas deserts. *(Boerhaavia coccinea).* Red-purple. May-September.

Four-o'clock family. Size: Trailing stems up to 4 feet in length.

A common roadside perennial, spiderling becomes an annoying garden weed when it invades open fields and areas of cultivation. Its trailing stems and sticky foliage interfere with tillage. The flowers are small but numerous and grow in attractive, colorful clusters. This species is widely distributed, not only throughout the deserts of the Southwest, but also in tropical and subtropical America.

In addition to *coccinea,* other species of *Boerhaavia* are widespread throughout areas of the Southwest below 5,500 feet elevation. The plants usually grow where they are exposed to full sunlight, although sometimes found in open brushlands, and reach full flower in late summer and autumn months.

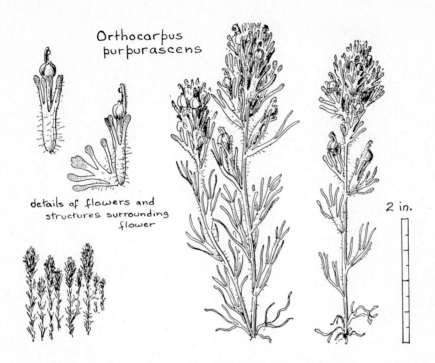

Orthocarpus purpurascens

details of flowers and structures surrounding flower

2 in.

Common Names: **OWLCLOVER, ESCOBITA OWLCLOVER**
Arizona and California deserts. *(Orthocarpus purpurascens).* Red-purple.
 March-May.
Figwort family. Size: 4 to 8 inches high.

This short, leafy annual ranging in color from rich velvet red to purple is noticeable even as an individual plant, but, following winters of above average rainfall, it often grows en masse, covering portions of the desert floor with a carpet of bright purple; some times in pure stands, often mixed with goldpoppy, lupine, and other spring flowers.

Since this species is limited in range to southern and western Arizona, California, and Lower California at elevations below 3,000 feet, Organ Pipe Cactus National Monument is well within its range, and in that area can be seen at its spectacular best.

The California variety has the lower lip of the blossom tipped with rich yellow.

PURPLE

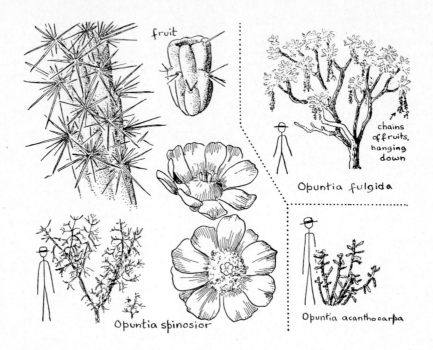

fruit

chains of fruits, hanging down

Opuntia fulgida

Opuntia spinosior

Opuntia acanthocarpa

Common Names: **CHOLLA, WALKINGSTICK CHOLLA, BUCKHORN CHOLLA (CANE CACTUS).**

Arizona desert. *(Opuntia spinosior)*. Red-purple. May-June.

California desert. *(Opuntia acanthocarpa)*. Yellow-purple. April-May.

Texas desert. *(Opuntia imbricata)*. Red-purple. May-June.

Cactus family. Size: Shrubby, from 3 to 8 feet high.

Aside from the tree-sized, Sonora jumping cholla *(Opuntia fulgida),* which is the largest of the branching, cylindrical-jointed cactuses and is very common in restricted portions of the desert in the Tucson-Phoenix area, the species listed above are the largest, most representative, and most widely spread of the chollas (CHOH-yahs).

The bright red to purple flowers of *Opuntia spinosior* and *Opuntia imbricata* make them particularly attractive during the blossoming season, while the extreme variability, from yellow to red and purple, of the flowers of *Opuntia acanthocarpa* make its identification by this means always a matter of uncertainty. Fruits of *spinosior* and *imbricata* are quite large, yellow, and at a distance may be mistaken for blossoms.

Flowers of *Opuntia fulgida* are small, pink, and appear in mid-summer followed by fruits which remain on the plant to form long hanging clusters relished by cattle. A hybrid between *spinosior* and *fulgida* is reported along the Gila River west of Florence, Arizona.

Cirsium
neomexicanum

Cirsium
undulatum

Common Names: **THISTLE, WAVELEAF THISTLE (MOJAVE THISTLE, NEW MEXICO THISTLE).**

Arizona desert. *(Cirsium neomexicanum)*. Pink-purple. March-September.
California desert. *(Cirsium mohavense)*. Pink-white. Summer.
Texas desert. *(Cirsium undulatum)*. Red-purple. October.
Sunflower family. Size: 2 to 4 feet tall, sometimes taller.

Sometimes called bullthistles, these biennials or perennials with spiny stems, prickly leaves, and heavy flower heads ranging in color from white to purple need no introduction to most people.

The Mojave thistle is the commonest form found in southern California, being abundant, sometimes in dense stands, in open gravelly valleys, on rocky slopes, or about alkaline seeps in the Mojave Desert. Range of the New Mexico thistle extends westward to the eastern borders of the Mojave Desert. The "Sierra" thistle, with white blossoms, occurs in Death Valley National Monument at elevations between 4,000 and 5,000 feet.

Navajo and Hopi Indians are reported to use the thistle plant for medicinal purposes.

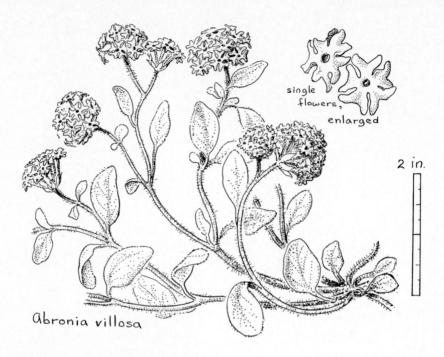

single flowers, enlarged

2 in.

Abronia villosa

Common Name: **SANDVERBENA**

Arizona and California deserts. *(Abronia villosa)*. Pink-purple. March-April.
Texas desert. *(Abronia angustifolia)*. Pink-purple. March-July.
Four-o'clock family. Size: Trailing annual, stems sometimes 2 feet in length.

Sandverbenas are attractive, low-growing herbs with pink-purple
to lavender, fragrant flowers forming clusters or heads which cover
the plants. Desert species are conspicuous in the springtime when
they line roadsides and carpet open, sandy locations, such as dry
streambeds, with a mass of purple. Although they are often found
in solid patches, they frequently intermingle with other spring
flowers such as the bladderpod producing a gay pattern of color.

Other species are found at higher elevations and are common
during the summer months.

Some of the desert species blossom a second time in September.

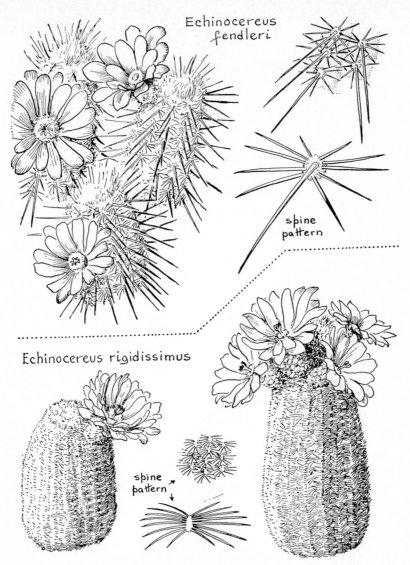

Echinocereus fendleri

spine pattern

Echinocereus rigidissimus

spine pattern

Common Names: **ECHINOCEREUS (HEDGEHOG CACTUS, STRAW-BERRY CACTUS, CALICO CACTUS).**

Arizona and California deserts. *(Echinocereus engelmannii).* Lavender-purple. March-April.
Texas desert. *(Echinocereus fendleri).* Pink-purple. May-June.
Cactus family. Size: 6 to 18 inches high.

Growing in open clumps with stems resembling spine-covered cucumbers standing on end, the hedgehog is the first cactus to blossom in the spring. Flowers vary considerably in color ranging from lavender through purple to a rich red.

Fruits (called "pitayas" in Texas) are dark mahogany red, juicy, rich in sugar, and may be eaten like strawberries, hence the name strawberry cactus. They form an important item in the diet of birds and rodents. Pima Indians consider them a delicacy.

PURPLE

A closely related cactus, the rainbow *(Echinocereus rigidissimus)* is restricted in its distribution to elevations between 4,000 to 6,000 feet. It is called "rainbow cactus" because of alternating bands of red and white spines encircling the stem and marking growth of different seasons and years. Its blossoms are pinkish (yellow in western Texas) and are large and showy in comparison with the small size of the single-stemmed plant.

Common Names: **PURPLEHEAD BRODIAEA (DESERT-HYACINTH, PAPAGOLILLY, BLUEDICKS, COVENA, GRASS-NUTS).**

Arizona and California deserts. *(Brodiaea pulchella).* Light blue. February-May.

Lily family. Size: About 1 foot high.

Very common and abundant in early spring, the pale blue to violet flowers of this small, delicate perennial lily are conspicuous on open slopes and mesas. Found below 5,000 feet from southwestern New Mexico to California and northward to Oregon, they are widely scattered over the desert areas of the Southwest. Pima and Papago Indians ate the small bulbs, as also did the early white settlers who named them grassnuts.

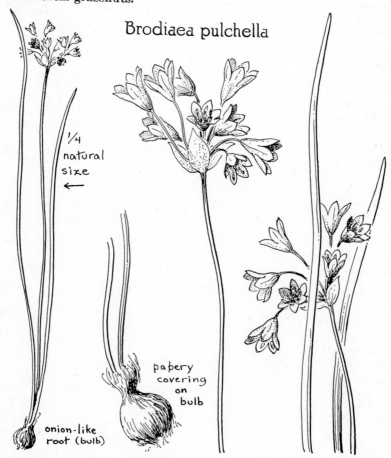

Brodiaea pulchella

¼ natural size ←

papery covering on bulb

onion-like root (bulb)

Common Name: **EVOLVULUS (WILD-MORNINGGLORY).**

Arizona desert. *(Evolvulus arizonicus).* Sky blue. April-October.
Texas desert. *(Evolvulus alsinoides).* Azure blue. April-September.
Convolvulus family. Size: Spreading perennial herbs up to 2 feet.

Although *Evolvulus arizonicus* is considered one of the desert's most beautiful wildflowers, members of the genus are by no means limited to the desert. They are found in sunny locations on desert grasslands, open plains and dry mesas below 5,000 feet from the Dakotas and Montana to Argentina.

The flowers, although rarely more than ½ inch in diameter, are bright azure or sky blue, and seem large in comparison with the small leaves and weak, spreading stems of the plant that bears them.

Although the genus *Ipomoea* is the true morningglory, blossoms of *Evolvulus* are similar in appearance, although flattened, hence are sometimes called wild-morningglory.

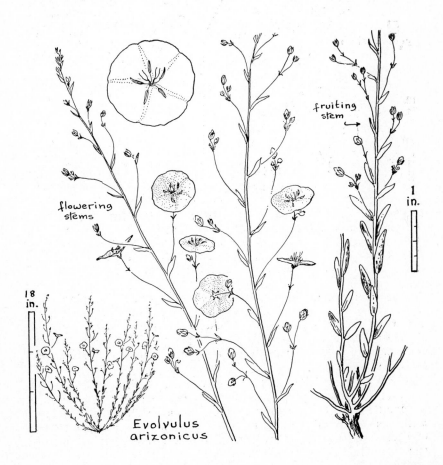

fruiting stem →

flowering stems

1 in.

18 in.

Evolvulus arizonicus

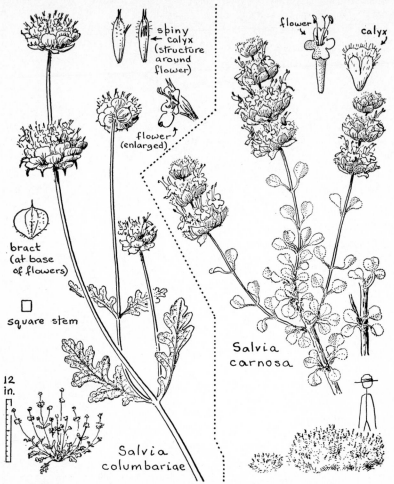

spiny calyx (structure around flower)

flower (enlarged)

bract (at base of flowers)

square stem

12 in.

flower

calyx

Salvia carnosa

Salvia columbariae

Common Names: **SAGE, DESERT SAGE, CALIFORNIA CHIA**
Arizona desert. *(Salvia carnosa)*. Sky-blue. Spring.
California desert. *(Salvia columbariae)*. Blue. March-April.
Texas desert. *(Salvia arizonica)*. Indigo blue. July-September.
Mint family. Size: Herbs and shrubs up to 3 feet high.

The word "sage" is derived from the idea that these plants had the power to make a person wise or sage. Please do not confuse the desert sage with sagebrush *(Artemisia)* which does not grow in low-elevation deserts but which, due to popular writings and "western" movies, is associated in the public mind with any brushy plant found in the west.

Seeds of the California chia once formed an important item in the diet of desert Indians and were used to remove particles of foreign material from their eyes. The seeds are still used by Mexican natives as food and for making mucilaginous poultices. The flowers of several species are very ornamental and the plants are quite common, usually in sandy soil.

BLUE   **100**

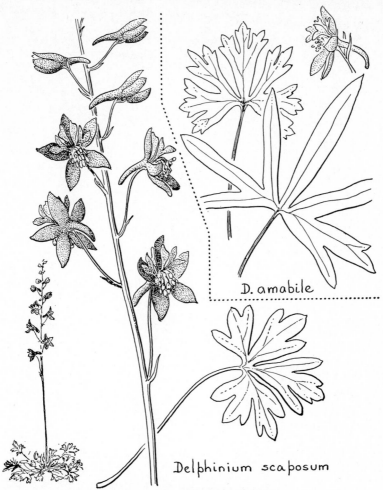

D. amabile

Delphinium scaposum

Common Names: **LARKSPUR, WILD-DELPHINIUM**
Arizona desert. *(Delphinium scaposum)*. Royal blue. March-May.
California desert. *(Delphinium parishii)*. Sky blue. Spring.
Texas desert. *(Delphinium carolinianum)*. Blue. Spring.
Crowfoot family. Size: Up to 2 feet in height.

Desert larkspurs are low-growing, spring or early summer flower-ing in habit, often occurring in colonies, and frequently intermingle with other spring flowers thereby adding their blue to the colorful tapestry of ground cover. They are readily recognized because of their resemblance to the cultivated varieties called delphiniums, and because of the tubular extension or "spur." *Delphinium amabile* is the most drought-resistant of all southwestern species and may blossom in the desert as early as February.

Because they contain delphinine and other toxic alkaloids, lark-spurs are poisonous to livestock, particularly sheep. On the desert, the plants are small and bear few but beautiful blossoms. They pre-fer open, gravelly soil.

It is reported that the Hopi Indians grind larkspur blossoms with corn to produce blue meal.

BLUE

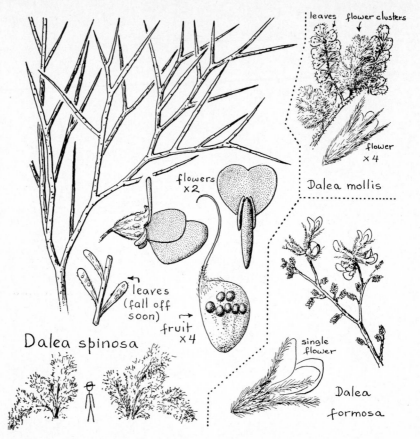

leaves flower clusters

flower
×4

Dalea mollis

flowers
×2

leaves
(fall off
soon)

fruit
×4

Dalea spinosa

single
flower

Dalea
formosa

Common Names: **DALEA, SMOKETHORN (INDIGOBUSH, PEABUSH).**
Arizona and California deserts. *(Dalea spinosa).* Blue-violet. April-June.
Texas deserts. *(Dalea formosa).* Purple. March-June.
Pea family. Size: Up to 10 or 12 feet tall.

Famous, although not common, throughout the frostless areas
of the desert, the smokethorn, because of its gray-green, leafless,
plume-like growth resembles at a distance a gray cloud of smoke
hovering over a desert campfire. When in flower, in May or June,
it is one of the handsomest of desert shrubs. It is always found in
the bed of a sandy wash where it obtains moisture from runoff
following summer showers or winter rains.

In California, it occurs in portions of both the Mojave and the
Colorado Deserts, and in Arizona is restricted to the western part
of the state. It is fairly abundant near Quitobaquito in the southwestern corner of Organ Pipe Cactus National Monument and in
Joshua Tree National Monument.

Other species of indigobush, of which there are many, are less
famous than the smokethorn, but all have purple or indigo flowers
and most of them are beautiful and noticeable when in blossom.
Indians used an extract from the twigs for dyeing basket material
and ate the roots of *Dalea terminalis.*

BLUE   102

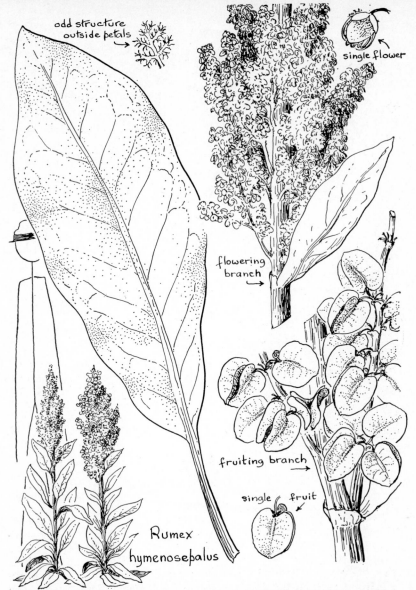

odd structure outside petals

single flower

flowering branch

fruiting branch

single fruit

Rumex hymenosepalus

Common Names: **DOCK, CANAIGRE (WILD-RHUBARB, SORREL).**
Arizona and California deserts. (*Rumex hymenosepalus*). Pink-green.
   March-April.
Texas desert. (*Rumex mexicanus*). Pink-green. Summer.
Buckwheat family. Size: Coarse perennial up to 2 feet tall.

    Sturdy, conspicuous flower and seed heads together with the
large leaves of these coarse, roadside plants, although hardly to be
considered as beautiful, attract considerable attention and arouse
the curiosity of the observer. Some species are garden weeds introduced from Europe. In the desert, the large, coarse leaves and pinkish flower stalks make quite a showing in sandy washes and along
the roadsides as early as March and April. The plant is being considered as a source of tannin (from its tubers) to replace that
formerly obtained from chestnuts.

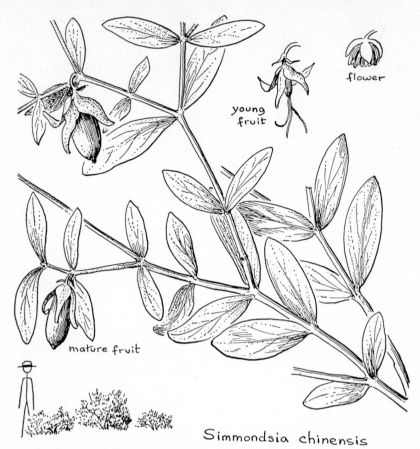

young fruit

flower

mature fruit

Simmondsia chinensis

Common Names: **JOJOBA (GOATNUT, DEERNUT, WILD-HAZEL, COFFEEBERRY).**

Arizona and California deserts. *(Simmondsia chinensis).* Green-yellow. December-July.

Box family. Size: Shrub, 2 to 5 feet high.

Jojoba (hoh-HOH-ba) is another of the desert plants which is noticeable, not because of its flowers, but due to its leathery, gray-green foliage which persists throughout the year. These shrubs are numerous at elevations between 1,000 and 4,300 feet in the lower levels of desert mountain ranges, particularly on the alluvial fans at the mouths of canyons.

The acorn-like nuts, which taste something like filberts, but are bitter because of their tannin content, were long an important item of food among the Indians and the early settlers. The thickly set, evergreen leaves are browsed by deer and other animals, and the nuts are gathered by ground squirrels.

The nuts contain an edible oil (actually a liquid wax) which has some medicinal value and is used in small quantities in the manufacture of hair oil. Attempts to raise the nut in commercial quantities have not proved successful. On occasions the nuts have been roasted and used as a substitute for coffee.

GREEN

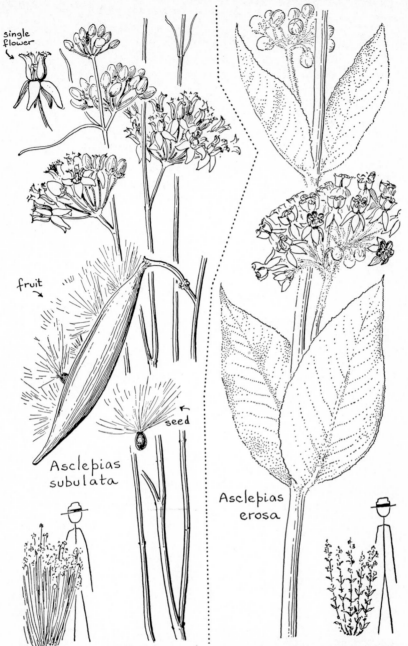

single flower

fruit

seed

Asclepias
subulata

Asclepias
erosa

Common Names: **MILKWEED, DESERT MILKWEED (BEDSTRAW MILKWEED, AJAMENTE).**

Arizona desert. *(Asclepias subulata).* Green-yellow. April-October.
California desert. *(Asclepias erosa).* Green-white. September-October.
Texas desert. *(Asclepias texana).* Green-white. Autumn.
Milkweed family. Size: Perennials, up to 5 feet.

Readily recognizable because of their milky sap and the pods filled with silky-winged seeds, the milkweeds are generally con-

GREEN

sidered poisonous to livestock, although rarely eaten. Appreciable quantities of rubber are found in the sap of some species.

Common Names: **ALLTHORN, HOLACANTHA (CRUCIFIXION-THORN, CORONA-DE-CRISTO, CROWN-OF-THORNS).**
Arizona desert. *(Koeberlinia spinosa)*. Greenish. May-June.
California desert. *(Holacantha emoryi)*. Yellow-green. June-July.
Texas desert. *(Koeberlinia spinosa)*. Greenish. May-June.
Koeberlinia is Junco family.
Holacantha is Simaruba family. Size: From 2 to 10 feet high.

Two intricately branched, thorny shrubs with green bark and leaves reduced to small scales and otherwise resembling each other are both popularly known by the names of crown-of-thorns and crucifixion-thorn, although they are not closely related botanically.

Flowers of both are small and inconspicuous, although when the bushes are in full bloom, they are quite noticeable. Even so, it is the unusual and eye-arresting appearance of these shrubs which appear

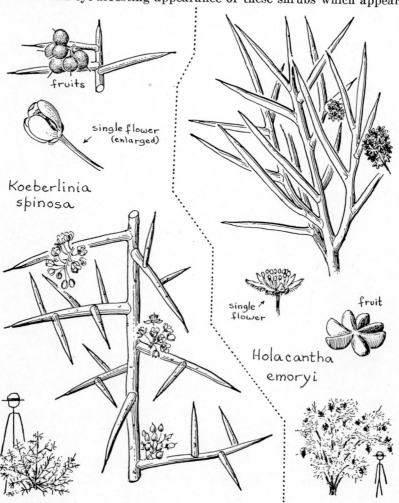

fruits

single flower
(enlarged)

*Koeberlinia
spinosa*

single
flower

fruit

*Holacantha
emoryi*

as leafless masses of robust thorns, making them a conspicuous feature of the desert and arousing the interest and curiosity of observers.

In some locations these shrubs are sufficiently abundant to form thickets which repel livestock. Fruits of *Holacantha emoryi* remain on the plant for years, and it is usually possible to identify each season's fruit clusters by the degree of weathering. These masses of brown to black fruits are very noticeable and are often mistaken for parasitic growth or the results of a disease. A somewhat similar shrub, sometimes attaining tree size and superficially resembling the paloverde (see p. 36), is the canotia *(Canotia holacantha)*. It is found at elevations between 2,500 and 4,500 feet, over much of southern and western Arizona and northern Sonora, and blossoms from May to August.

Common Names: **BURSAGE (BURROWEED, BURROBUSH).**
Arizona desert. *(Franseria deltoidea)*. Greenish. December-April.
California desert. *(Franseria dumosa)*. Greenish. April-November.
Sunflower family. Size: Up to 3 feet high.

Noticeable because of its ashy foliage, bursage is a low, rounded, white-barked shrub, the several species of which are very common on the dry plains and mesas up to 3,000 feet. The flowers are small, without petals, and colorless inasmuch as they are wind-pollinated and do not need to attract insects.

It is classed by A. A. Nichols as one of the major plants of the paloverde-bursage-cactus plant association, one of the three plant communities of the Sonoran Desert.

Bursage is one of the favorite foods of burros and sheep, and is said to be preferred also by horses.

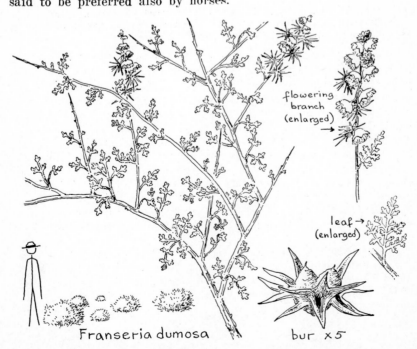

flowering
branch
(enlarged)

leaf →
(enlarged)

Franseria dumosa   bur x5

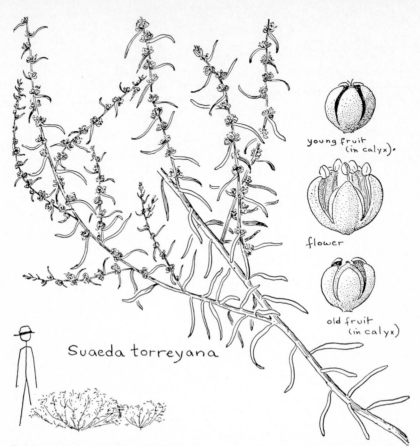

young fruit (in calyx).

flower

old fruit (in calyx)

Suaeda torreyana

Common Names: **SEEPWEED (INKWEED, IODINEBUSH, QUELITE-SALADO).**
Arizona and Texas deserts. *(Suaeda suffrutescens)*. Greeenish. March-July.
California desert. *(Suaeda torreyana)*. Greenish. July-September.
Goosefoot family. Size: Up to 6 feet tall, and branching.

Seepweed, which is usually an indicator of alkaline soil, is browsed to some extent by cattle when other feed is scarce. The young plants are raised for greens by the Pimas and other desert Indians, sometimes eaten with cactus fruits. *Pinole* was made by roasting the seeds. Coahuila Indians extracted from the plants a black dye which they used in artwork.

Flowers of the seepweed are small, greenish, and without petals. Since the pollen is carried by the wind, color to attract insects to the flowers is not necessary. Because of its tolerance for somewhat salty or alkaline soils, seepweed thrives along the margins of dry lakes and on salt flats where moisture is near the surface. On the desert of southern California it is often associated with mesquite and quailbrush, the sooty-green to brown plants standing out in sharp contrast.

Because it is so common in moist locations through the Southwest, and sufficiently unusual in appearance to arouse curiosity as to its identity, seepweed is included in this publication regardless of the fact that its flowers are small and inconspicuous.

GREEN

108

INDEX

110

111